T0073599

# THE WEAKEST LINK

# THE WEAKEST LINK

**How to Diagnose, Detect, and Defend Users from Phishing**

ARUN VISHWANATH

**The MIT Press**
**Cambridge, Massachusetts**
**London, England**

The MIT Press would like to thank the anonymous peer reviewers who provided comments on drafts of this book. The generous work of academic experts is essential for establishing the authority and quality of our publications. We acknowledge with gratitude the contributions of these otherwise uncredited readers.

This book was set in Adobe Garamond Pro by Westchester Publishing Services. Printed and bound in the United States of America.

Library of Congress Cataloging-in-Publication Data

Names: Vishwanath, Arun, author.
Title: The weakest link : how to diagnose, detect, and defend users from
    phishing / Arun Vishwanath.
Description: Cambridge, Massachusetts : The MIT Press, [2022] | Includes
    bibliographical references.
Identifiers: LCCN 2021060548 (print) | LCCN 2021060549 (ebook) |
    ISBN 9780262047494 (hardcover) | ISBN 9780262371964 (pdf) |
    ISBN 9780262371971 (epub)
Subjects: LCSH: Phishing. | Computer security. | Computer networks—
    Security measures. | Computer crimes—Prevention.
Classification: LCC HV6773.15.P45 V56 2022  (print) | LCC HV6773.15.P45
    (ebook) | DDC 364.16/8—dc23/eng/20220307
LC record available at https://lccn.loc.gov/2021060548
LC ebook record available at https://lccn.loc.gov/2021060549

10   9   8   7   6   5   4   3   2   1

# Contents

# Acknowledgments

This book is a culmination of a journey that began with a spear phishing attack my institution received more than a decade ago. It was a new form of attack for the time, different from the all-to-common Nigerian phishing email. The attack occurred when I was studying the psychology of technology use and testing different ways of framing messages to persuade users into using their devices optimally. It was this fortuitous timing that led me to recognize the potential of this new attack vector, to my overall body of research on user risk from phishing, and ultimately to this book.

The research journey that followed took years of trying and failing. Many helped along the way. Among them were numerous students who worked on my research. Some worked for course credit, others coauthored papers with me, still others served as subjects, volunteering their data. This book wouldn't have been possible without their contributions. There were also agencies and organizations that lent invaluable support. The National Science Foundation provided some of the initial funding. Other organizations allowed me to test my approach on their employees. They provided data and insights into the challenges they faced, which guided the development of the cyber risk assessment approach. I am thankful to each of them.

There are many others. Most notably, Bruce Schneier at the Harvard Kennedy School, who inspired and mentored me through the arduous book writing process, and Jeff Dean, my former editor at Harvard University Press, who was among the first to see the book's potential. I am forever grateful to them. I am also thankful to the entire editorial team at the MIT Press, who stepped up and shepherded the book through the publication process. Others include Dr. Loo Seng Neo, formerly at the Singapore Ministry of Home Ministry (and the rest of the behavioral science research team headed by Dr. Majeed Khader), and Simon Pavitt, at the UK Ministry of Defence, who helped refine the cyber hygiene inventory. They, along with the community of national security and law enforcement professionals in the United States, the United Kingdom, Europe, and Australia, helped shape this book, and I am grateful to all of them.

My final, deepest gratitude is to my family. My wife, Leslie, and my children, Vera and Dean, patiently suffered through my years of research and writing. Without their support, the book wouldn't have happened. Without them, none of it would matter. Thank you!

# INTRODUCTION

It was Monday, November 24, 2014. Employees logging into their computers were greeted by a locked screen, across which flashed a menacing image of a fiery red skull with long tentacles with the message "Hacked by #GOP." Accompanying it were sounds of gunfire, a poorly worded warning alluding to the theft of all the company's internal data, and a deadline of 16 hours to comply with a demand. This was the beginning of a hostage situation—one that would rewrite the rules of cyber warfare forever.

The company under siege was Sony Pictures Entertainment (SPE). A hacker group named GOP, short for Guardians of Peace, demanded the stoppage of *The Interview*, a Seth Rogen movie slated for release on Christmas Day that featured a comical plot to assassinate North Korean leader Kim Jong-un.

As harried SPE employees restarted their computers, the malware kept finding newer hosts, quickly leaping from computer to computer, jumping through networks and then through servers. Within an hour, the attack had infected all SPE computers in Los Angeles, then New York, and soon across all continents. Within a few more hours, everything digital—files, data, emails, messages, scripts, storyboards had been irretrievably lost.

Writing for *Fortune* magazine, Peter Elkind detailed the scale of the destruction: "It erased everything stored on 3262 of the company's 6797 personal computers and 837 of its 1555 servers. To make sure nothing could be recovered, the attackers had even added a little extra poison: a special deleting algorithm that overwrote the data seven different ways. When this was done, the code capped each computer's startup software, rendering the machines brain-dead."[1]

Over the next few weeks, the hackers dumped batches of confidential files on publicly accessible file-sharing hubs. These included emails among SPE's leadership team, the salaries and social security numbers of 47,000 employees, passports and visas of various cast and crew members, unfinished and finished scripts of yet-to-be-released movies, and even information about SPE's corporate vendors, such as the salary data of over 30,000 employees of Deloitte, its accounting firm. In all, hackers stole and released over 100 terabytes of data.

By mid-December, the Federal Bureau of Investigation (FBI) had officially attributed the breach to North Korea. The overall cost for system cleanup and recovery would be a staggering $45 million. That's without accounting for the firing of SPE's studio chief, Amy Pascal, and others in top management; the loss of revenue from the leaked movies and scripts; the class-action lawsuits from employees and vendors; and the months of embarrassment from the trove of confidential emails that revealed not just the insides of the movie business but also SPE executives' antipathy toward President Barack Obama and various Hollywood stars and starlets.

But while the media were busy covering the salacious gossip, there was a critical question no one asked: how could a country like North Korea pull off such a major cyber breach? To put this in context, all of North Korea's 24 million inhabitants have access to just about 28 websites, and only 0.3 percent of its entire population—7,200 people—have unrestricted web access.[2] In contrast, SPE's parent company, the Sony Group, employs 114,000 people around the world, all of whom enjoy unrestricted web access. Even by itself, SPE's revenue of $8 billion is more than the combined import and export revenue of North Korea.[3] So how did this technologically unsophisticated nation push one of the world's foremost technological corporations back to the precomputer age, where employees were now resorting to Post-It notes and bulletin boards for communication? The answer is spear phishing—a virulent, internet-based social engineering attack that I had been tracking, researching, and warning about for almost a decade.

Spear phishing uses old-school confidence tricks to get unsuspecting users to provide their online account passwords on fake websites or click on malicious hyperlinks and attachments that provide direct access to their device. Attacks are deployed via email but can just as easily be sent using social media, text messages, USB storage devices, Wi-Fi networks, or Bluetooth connections.

Using this access, a hacker can lurk on a victim's computer, virtually move within an organization's computer networks, and install destructive programs to siphon off data or prevent access to it. Basic programming skills and a free email account for deployment are all it takes to craft one of these attacks, and, since the attacker takes on the virtual identity of the system's approved users, there is no way to identify and block them once they enter the system.

Sony, like just about every other company in the world, provided all its employees with email and internet access. This made spear phishing the perfect attack vector for a less technologically developed adversary such as North Korea. They needed a free email account and needed malware concealable in a hyperlink. One outgoing internet connection could procure this. Using it, they could craft the attack and send it to everyone at Sony—and they could do this repeatedly. All it would take to start the breach was for just one of the thousands of SPE's email users targeted to click on the malware. Email users were the conduit for the breach—they were the weakest link.

The SPE attack was a game changer because it showed how spear phishing could easily co-opt users and wreak havoc on organizations. Thanks to the ubiquity of emails, this virtual surface area for exploit within any organization's network infrastructure was vast and easy to access. Now that the hackers who launched the SPE attack had exposed this potential, hackers everywhere began to copy them. Weak links everywhere were being targeted, and no organization was safe anymore.

## SPEAR PHISHING ATTACKS EVERYWHERE

Not long after the SPE attack, in February 2015, came news of another major breach. Hackers had stolen the personal health data of 80 million customers of Anthem, Inc., the nation's largest health insurance provider.

Close on its heels, in June, news media reported an even bigger breach. This time, Chinese military hackers had exfiltrated 21.5 million user records from the Office of Personnel Management (OPM)—the agency that manages human resources and conducts security clearances for all US government employees. Many considered it the biggest breach of sensitive federal government data in modern American history. It was.[4]

In July 2015 came news of an attack on adult dating website Ashley Madison. This time, a hacker group calling itself The Impact Team had stolen 25 gigabytes of data from the site's servers, including the personal details of 37 million users, 13 gigabytes of corporate emails, and all company financial records. The hackers held the website's owner, Avid Life Media (ALM), hostage, demanding they shut down Ashley Madison and its sister website Established Men. When this demand wasn't met, the hackers released all the stolen data on a public file-sharing site.

As with Sony, the fallout caused months of public embarrassment for ALM, resignations of many key personnel, millions of dollars in cleanup costs, and the payout for a major class-action lawsuit. Sadly, a few embarrassed users committed suicide. The virtual attack had now taken real lives.[5]

News of attacks soon started pouring in from all over the world. By 2017, spear phishing had become the leading threat to cybersecurity globally, responsible for 93 percent of all breaches. It was the attack vector for everything from gathering information to installing malware, key loggers, and backdoors to exfiltrate data,[6] and it was being used by activist hackers ("hacktivists"), disgruntled employees, terrorist groups, organized crime networks, and state-sponsored espionage rings from Russia, China, Iran, and other nations.

Within a few years after the SPE attack, a spear phishing attack had targeted every person in the world with an email account.[7] Many had already become a victim of some breach—most just weren't aware of it yet.

The situation was even direr for organizations. Many were constantly under attack, and the sheer ferocity of the attacks was astonishing. In a meeting I attended, the US Senate's sergeant at arms reported blocking 23 million emails in 2017; the US House of Representatives reported blocking millions more. That year, the Department of Defense received 36 million malware-laden emails per day, or 13 billion emails.[8] The responsibility for stopping this deluge fell on information technology (IT) managers—on whose shoulders it has remained.

## THE "PEOPLE PROBLEM"

IT managers all over the world have taken one of two approaches. The first, rooted in the engineering tradition, focuses on using technology to thwart attacks. The goal here is to reduce the quantity of spear phishing emails that reach users and, if the emails do reach anyone, restrict the damage.

The sheer volume of email-based exchanges across all the many devices people use limits this approach. It is difficult for IT to read all online communication, monitor all devices, and effectively limit all spear phishing emails from getting to users without invading their privacy. The other limitation comes from the speed of current computing innovations, which changes so rapidly that proactively finding the weaknesses in each new technology—let alone keeping the virtual hatches on all of them tightly latched to forestall all breaches—is impossible.

This realization has led to the second approach: "hardening" people. Organizations harden their employees, or make them more resistant to spear phishing, by using different awareness training approaches that aim to educate users and improve their ability to detect the deceptive clues in the emails.

Most cybersecurity training takes one of two forms. One is a type of didactic training that involves in-depth, focused educational sessions conducted online or in classroom-type settings. Most such training uses an exam or test after the training to certify the end user's readiness. The second form of training involves subjecting employees to simulated spear phishing attacks (called pen tests) followed by a short educational module embedded in the attack. In most such "embedded training programs," the educational module, provided only to the users who fail the test, reveals the deception clues they missed in the email. The assumption here is that being deceived shocks users and makes them more receptive to the training that follows.

Today, driven by the fear of getting hacked, almost every organization in the US—and increasingly all over the world—deploys one or both forms of training. For many federal and state government employees and people working in specific sectors, completing cybersecurity training is mandatory, so organizations expend significant resources not just in actual expenses but also in time spent on training.

Objective data on training effects paints a bleak picture, however, showing training as either having no effect or having merely a short-term influence on user behavior. In academic research experiments, within days and at times a mere few hours after being trained, users revert to their established patterns of email use and fall victim to the very same phishing attacks they were trained to detect.[9] Other evidence comes from my own interactions with organizational leaders who have privately shared their frustrations with the limited efficacy of training on their users. Perhaps the best evidence comes from continued news stories of breaches in technology companies (e.g., Adobe in 2013, Deloitte in 2017, the Financial Services Information Sharing and Analysis Center in 2018, and Microsoft in every year since at least 2019), where users are not just highly trained but also technologically adept. So, even users in companies that people turn to for advice on spear phishing fall victim to it.

Faced with such data, the usual reaction from IT leaders in organizations is to train their users even more. Whenever training fails, organizations resort to even more training. This is because everyone in IT believes that training works. Most think it is so effective that they are okay with punishing users who fail training tests. Admiral Michael Rogers, former head of the US National Security Agency (NSA), famously stated the need to court-martial users who fail spear phishing tests.[10] While most organizations don't go that far, many use punishments ranging from mild warnings to users being reprimanded, singled out, or humiliated for their failure. This strategy of shaming and flaming usually leads to even more training being inflicted on the user, and if none of this improves spear phishing awareness among employees—which happens more often than admitted—IT managers usually end up blaming their users for being untrainable. They call it the "people problem," a catch-all term for everything users do that can lead to security breaches in organizations.

But it gets worse. Not only is training today blindly prescribed to treat the people problem of security, IT managers also use it as a metric of the overall cyber readiness of an organization. The pressure for this comes from regulators and policymakers, as well as from cybersecurity insurers, who have been asking for quantitative cyber risk metrics for their actuarial computations. To fuel more sales, many training companies also advocate this approach. As a result, organizations increasingly use their user training data—pass or fail on didactic training or failure rates on embedded phishing pen tests—as a metric of user cyber risk in their organization. This compromises an already flawed approach in at least five important ways.

For one thing, the current paradigm of using pen-test success or failure rates as a diagnostic for cyber risk conflates outcomes with causes. Failing a phishing test is an outcome, no different from having an automobile accident. Just as an automobile accident can happen because of factors within the control of the driver (e.g., speed, attentiveness, quality of car) and outside factors (e.g., quality of road signage, other drivers), failing a spear phishing test is also merely an outcome caused by many factors.

Second, if our goal was to reduce accident rates in a group of accident-prone individuals, the last thing we would do is try to teach them by inflicting an accident on them and punishing them for it. But that is precisely what all the many phishing tests do: they find crafty ways to make users fail a test and then accord punitive treatments for failing. If this were our approach to driver's ed, it would lead to more paranoid drivers rather than safer ones.

Third, if our goal was to improve how accident-prone drivers could drive better, our immediate assumption wouldn't be that they all simply lack safe-driving awareness. After all, some in our group might have had accidents because of poor eyesight, others may be driving impaired from some substance in their system, and still others might be inattentive because talking or texting while driving distracted them. Simply imparting more-rigorous knowledge wouldn't help those for whom knowledge or awareness wasn't a reason for their poor driving. Likewise, repeatedly conducting awareness training on everyone by using the same approach does little to protect the organization if the underlying reasons for phishing susceptibility vary by user.

Fourth, if our goal is to make users capable of safely driving a single car on a stretch of road under the most predictable circumstances, then perhaps having them comply with the same limited set of instructions is all we need. But driving is far more complex and requires adaptations based on the type of car, traffic, road surface, and weather. Auto safety requires some compliance, such as always wearing seatbelts, but also requires a good degree of "internalization"—a social science term that means incorporating the information into one's consciousness and using it as a principle to guide decision-making. It also needs good driving habits such as routinely checking rearview mirrors, looking over your shoulder when merging, and signaling before turning. Instilling all this requires more than training. It requires imparting knowledge, fostering motivation, and nurturing good driving habits.

Compared to roadways, the information superhighway is vaster and more unstructured. Cyberattacks are not just more frequent but also more unpredictable, because they mutate with new software. Some vulnerabilities (known as "zero days") aren't even known until a hacker exploits them. Yet our approach to cybersafety is less developed. We train everyone under the singular assumption that they lack awareness. Everyone gets the same awareness knowledge, regardless of its relevance or fit to their needs. Everyone is then subjected to even more testing—which, when many users invariably fail, leads to more of the same training. Nowhere in this process is there any measurement of what users needed, what they lacked, or why they failed. The outcome would be just as poor with automobile safety if we failed to consider a driver's vision problems and instead focused on repeatedly teaching them how to read road signage.

Finally, the goal of auto safety is to make not just one driver safe but to make the entire system safe. A big impetus for this has been the auto insurance system, which rewards and punishes drivers for safe driving. But the system isn't arbitrary. There are insurance adjusters and law enforcement personnel who investigate the cause of accidents, interview drivers, and get feedback. The system allows access to this information, creating accountability. It also gives feedback, allowing users to reduce premiums, giving them agency over their points. In turn, this has driven demand for automobiles with more safety features, for manufacturers to innovate, and for users to be more careful while driving—which together have made automobile safety the norm rather than the exception.

A similar approach is missing from our present regimen for training users in cybersecurity. While we use different phishing simulations and score users who pass or fail, we punish those who fail, usually with even more training, and ignore those who pass. Today, insurance companies indemnify organizations from cyber losses, but they, too, rely on the training data provided by the organization's IT department to decide on premiums.

The problem is that no one really knows, or cares to know, why the user passed or failed the test—not the person in IT crafting it or even the user, who is neither asked about nor understands their performance. So, it's unclear whether the test was flawed, whether passing or failing was a fluke, or whether the outcome resulted from some conscious or unconscious thought or habitual action of the user. Because of this, it's hard to trust the scores from such training, which does little to build accountability or give users a sense of agency regarding their security posture.

Building a culture of cybersafety requires that we make users accountable for security—their own and that of the entire organization. This means more than simply scoring pen-test failures. It means understanding the thoughts and actions of users that led to their failure or success, the likelihood of them doing so again, and the likelihood that others would have fared similarly. It means receiving feedback both about the test and about the user. It also means giving users feedback, along with a road map for improving their security posture. If done consistently, this can help build user trust in the scoring process, create accountability, and foster a culture of cybersafety. Doing so will help make not just one user safe but everyone, from others in the organization to those the user takes this learning to—vendors, customers, and even people in their household. This is how cybersafety can change from being a dictate enforced by IT managers to a cultural and societal reality.

Achieving this requires an approach that can accurately capture the thoughts, actions, and habits that lead to user risk. Such an approach needs to be agnostic to the operating system and other software they are using, and the work they are doing. This way, we can assess and track the cyber risk of anyone, anywhere, and prepare them for present, potential, and even zero-day attacks. The approach also requires a risk classification system that is simple for all users to not just use but also comprehend. This way, we can tell them about the cyber risk they pose and help them reduce it.

For organizations, knowing the causes of user risk can lead to more-effective policies. Knowing why some users among them weren't at risk, meaning what some did right—a factor ignored in all our current approaches—can lead to novel solutions or interventions that can be learned, and this knowledge can be built on. Because such solutions emanate organically from within the user population, they can address the intraorganizational causes for user risk in ways no vendor-based training approach developed from outside ever can. This, by default, ensures that policies and security postures align with how users perform their work-related tasks. It further ensures that users won't find workarounds for policies that stand in the way of work, so IT staff can police less. Organizations can then reallocate these security resources and, for a fraction of the effort and cost, achieve cyber resilience.

Implementing such a risk assessment approach requires a method that is reliable, transparent, and compatible with what's already being done presently. This way, any organization, whether on Main Street or Wall Street, can use it. This risk classification system and scoring method is what this book provides.

## A DIAGNOSTIC FOR IDENTIFYING THE WEAKEST LINK

Conducting a cyber risk assessment that meets all the above criteria may appear daunting, but it doesn't have to be. Over the course of a decade, my research on technology users has uncovered the reasons some people fall victim to spear phishing and some don't. It has identified the metric and developed the method for measuring these reasons.

My research used simulated social engineering attacks on different users and figured out *how* and *what* people think that makes them susceptible or resilient. It developed a method for capturing these thoughts and behaviors reliably.

This work led to an important realization: there were less than a handful of user-level factors that led to deception or its detection. By measuring just these, any organization could pinpoint who among their users posed a risk to the enterprise and understand why.

But this approach faced an important limitation in that it required 40 or more survey questions. The questions were also complex, some involving terms that only users with a command of English would understand. Employees in many organizational settings couldn't or wouldn't be able to answer the questions.

My focus shifted to the development of a practical, globally usable measure. The need was for an efficient measure that was easy to understand and answer by anyone, in any organization, anywhere, within a few seconds. Through another series of research studies, this quest culminated in the identification of two simple questions, the responses to which were just as valid. These two questions and the approach for asking them is what I call the Cyber Risk Survey (CRS).

The CRS follows a phishing pen test, which many organizations already do as part of their training protocols. But, unlike the tests that are presently being done, it uses a defined framework for crafting the email and deploying it. I developed this framework, called the V-Triad, in 2017 and presented it at Black Hat USA. (The full presentation is available on You-Tube.)[11] In 2020, researchers from the National Institute of Standards and Technology (NIST) used the same approach I'd used and came to similar conclusions about phishing pen tests—independently underscoring the validity of the V-Triad.[12]

But the CRS achieves more than simply scoring pen-test quality. It helps design the optimal phishing pen test, collect feedback about it, deploy it on users, and capture users' susceptibility to it. The measures help in understanding who is at risk now as well as who is likely to be at risk in the future. This enables IT managers to pinpoint how much risk each user poses relative to others and locate the weak links among users.

The CRS is also underpinned by another framework: the Suspicion, Cognition, Automaticity Model (SCAM). Developed by my research team in 2016, this model comprehensively explains the user thoughts and actions that lead to social engineering deception or its detection. The model is built on years of testing on users who were subjected to various types of simulated social engineering attacks.

The SCAM provides the explanatory framework for the CRS. Using it to guide analysis, IT managers can go beyond simply locating the weak links; they can now explain why the links are weak. They can also identify the lower-risk users and understand the specific thoughts and actions that attenuated their risk. They can use this information to craft best practices that are particularly suited for the organization and face little resistance from users.

Explanations of the causes of user risk can be communicated with users. This can create a shared understanding of security issues, an internalization of solutions, and a culture of cyber safety. Last but not least, the CRS provides a net resilience score (NRS) for the overall organization. This score serves as a metric of organizational preparedness against future social engineering attacks, which can be used to compare the performance of organizational groups and for tracking improvements in their resilience over time.

Thus, the CRS is a comprehensive approach to the threat of social engineering. IT managers no longer need to worry about the adequacy of their user-focused approaches or wonder which user in the organization poses the biggest risk to the organization's cybersecurity. They can use the CRS along with its associated metrics to identify risky users, create accountability, craft policies and protections based on these results, and develop a culture of cybersafety. Using the information in this book, IT managers can solve their people problem of social engineering.

## THE GOAL OF THIS BOOK

This book arms the reader with the knowledge necessary for detecting, diagnosing, and defending against cyberattack by using the Cyber Risk Survey approach. It details the steps for implementing the approach, from how IT should conduct the ideal phishing pen test to how they should administer the CRS and generate the net resilience score.

Generating a valid NRS requires that managers understand social and cognitive science, empirical research, and approaches to analyzing behavioral data. It also requires understanding the fundamentals of social engineering and pen-testing development. This book covers all these topics. It provides a working knowledge of cognitive and behavioral science theories that explain how and why people get deceived via social engineering. It presents the CRS questions, the SCAM framework for analyzing responses to the questions, and the mechanism for converting user responses to the pen test into an NRS. It also covers the V-Triad and teaches how to craft the ideal phishing pen test and assess its quality.

This book further delves into cyber hygiene—a term that is widely used in security circles but is poorly understood—explicating what cyber hygiene means, what it entails, and what it doesn't. It presents the Cyber Hygiene Inventory (CHI), a scientifically developed tool for quantitatively measuring a user's cyber hygiene. It provides the inventory and discusses how the CHI and CRS can work together to quantify user risk. Finally, the book presents case studies of how different organizations used the CRS and CHI to achieve cyber resilience. In short, the book equips the reader with all tools for solving the problem of social engineering.

## WHY I WROTE THIS BOOK

My single most important driver for writing this book was frustration. You see, the November 2014 ransomware attack on SPE was not just a watershed moment in cybersecurity history. It also marked a turning point in my career. Until then, I had been a social-behavioral scientist studying users and how social engineering attacks deceived them. This followed years exploring the psychology of users and what drove them to adopt and use technology.

I'd been drawn to the study of social engineering by a series of spear phishing attacks that targeted my university's email users in 2009. The attacker asked users to update their passwords or risk losing access to their emails—a common attack today but less so back then. It was different from the typical poorly worded "Nigerian Prince" emails. The attacker focused on what mattered: a relevant subject line, a bold warning, a short deadline, and a highlighted hyperlink for changing a password.

The subject line is what most people viewed first; its relevance hooked them. The warning and deadline did the rest. Grammar and spelling errors were there, but the threat of losing access to their emails caused everyone to ignore such flaws and reeled them in. They hurriedly clicked on the centrally placed blue hyperlink, which took them to another official-looking website that captured their passwords. The attacker had crafted the email to take advantage of how people's minds worked online—it was designed by someone who understood user psychology and how to take advantage of it.

No one really noticed much of this or cared: not the users, most of whom complied, or even those in the IT department, who simply asked users to change their password—with another email, of course—and moved on. The problem was being solved using the same approach that caused it, a "solution" that many organizations (such as retail chain Target and credit-monitoring agency Equifax) implemented years later, with dire consequences.

Back in 2009, I was testing different content-based persuasive approaches on users, looking for ways to make them optimally use email, social media, and mobile devices to harness their fullest potential. The spear phishing attacks appeared to have a similar focus but with a darker agenda. They, too, were using content and other persuasive techniques to make online users do something—open, click, enable, accept—except their goal was to co-opt the user and misutilize the access.

I began by surveying the victims. I wanted to understand how they thought, the actions they took, and the impact their actions had. The attack's overwhelming success soon stood clear. Not only had most users complied, but none were even aware that they had been deceived. Many had acted without thinking, and those who had paused to think didn't have the know-how to correct their decision. Everyone had either thoughtfully or inadvertently complied. Most were victimized. This new form of attack was the ultimate vector of online deception. Anything could be concealed behind a hyperlink, even within an email, in an attachment. From there, the only limiting factor would be the motivations of the hacker.

My research focus completely shifted to examining this new vector. Much of my early work focused on replicating attacks reported in the media. I read about attackers using social media to impersonate people, using different appeals to elicit a response, using different devices (e.g., USB sticks, texting apps) to entice usage, and spoofing or mimicking different web pages to deceive users. After procuring the requisite approvals for studying human subjects, I experimentally re-created many of them. The goal was not just to examine how each attack succeeded but also to identify the specific facets that caused deception.

These tests initially used students in the US and abroad, then employees in organizations that had volunteered. With each study, I improved my approach, iteratively refined my measurement techniques, and better understood the problem. Each study gave me a better understanding of what hackers were doing, how their attacks were evolving, and how vulnerable organizations really were to this new form of attack.

By 2014, I was among the few researchers who had extensively studied the phenomenon of spear phishing from the user deception perspective. Thus, when the infamous Sony attack occurred in November of that year, I not only knew what the North Koreans had done but also exactly how they had accomplished it. I also knew what was coming—now that the proverbial genie was out of the bottle.

But the media completely ignored this. They focused on the salacious, inside Hollywood gossip: on what the emails said, not what the attack signified. I wanted to bring attention to what really mattered—the spear phishing attack, why it had succeeded, and the looming threat of more like it.

To this end, in December 2014, I wrote my first CNN opinion piece discussing how the North Koreans had accomplished the SPE breach by compromising users and what organizations and policy makers must do to stop such attacks. It would be the first of many more such attacks, because, as I'd expected, hackers all over the world started launching even bigger, bolder, and more consequential attacks.

So, I followed up with even more media pieces. Whereas once I had focused solely on addressing the research community, I now focused on communicating directly with cybersecurity professionals, technology users, IT managers, and policy makers. I became a technologist writing in the public interest. My goal was to bring attention to the core user-focused problems in cybersecurity and offer solutions to them. My hope was that by drawing attention to them, organizations and policy makers would attend to them before they mushroomed and became intractable.

I was among the first to call for user cyber hygiene—a term that has since become common in the field—and the first to develop an approach for measuring it. I advocated the implementation of two-factor authentication as a default setting, and called out the dangers of single sign-ons and fake social media profiles. I called for a nationwide system to forewarn others of cyberattacks, for privacy apps to have better "warning labels" and five-star rating systems, and for us to be concerned about the rise of mobile-based attacks, internet-of-things (IoT), and artificial intelligence. These ideas were featured on CNN and in *Wired* magazine, the *Washington Post*, *Politico*, *USA Today*, the *Chicago Tribune*, *Scientific American*, and other leading outlets.

I became one of the foremost experts on human cyber vulnerability. I was invited to present my research at the Coalition for National Science Funding (C-NSF) on Capitol Hill, multiple times at the US Senate and House by the Senate sergeant at arms, and at leading security conferences such as Black Hat. I presented at the US Army Cyber Institute at West Point, New York; at Johns Hopkins University; and to the cybersecurity thought leaders at the NSA, FBI, Department of Homeland Security (DHS), and the Obama White House's Office of Science Technology and Policy (OSTP). But despite the coverage, the publications, and the presentations, the problem of social engineering at its core remained. Nothing really changed. Attacks kept happening, becoming more frequent and significant.

Many of the problems I'd exposed were ignored—only to become even bigger in the years to come. For instance, after a spate of attacks on US hospitals, in 2016 I wrote about the rise of ransomware, calling it the year of the ransomware. I was wrong. It wasn't just in 2016; it's been every year since then. In 2021, we witnessed some of the biggest ransomware heists: Colonial Pipeline, which supplied half of the US East Coast's gasoline; JBS, the world's biggest meat supplier; and Kaseya, a supplier of remote monitoring software to IT companies that serves over one thousand small- and medium-sized businesses all over the world. What had been ransom demands of a few hundred dollars in 2016 are now up to $70 million in the case of Kaseya. What could have been stopped then is now going to cost millions more to clean up.

All the more frustrating was that the solutions I'd offered all along could have prevented much of this. I had presented a diagnostic for assessing users' cyber risk beliefs, their cognitive-behavioral schemas, and their habits; I'd developed a mechanism for using this to train users; and I'd even designed a tool for measuring cyber hygiene and developed an approach for conducting cyber hygiene assessments, all of which were publicly available.

Most security companies chose to ignore this. They doubled down on their own products, which they had already spent millions developing. Security researchers did the opposite. They took to rebottling my suggestions. Some relabeled risk belief as overconfidence, others called it user curiosity, while still others termed it the users' lack of technical efficacy. It's a phenomenon best described by Mark Twain as the old habit of calling everything new. This habit is endemic in academia, where scholars are rated on originality, not substance or value, so every academic is motivated to reinvent, rename, and reframe a problem rather than test or improve any existing one. Adding to this is the low stake in their pursuit. Neither a breach nor a data loss or its fallout matters to them personally or professionally.

But it does to IT managers and organizations the world over. They are at the mercy of vendors at one end, regulators at the other, and their customers throughout. They are caught between technologies that they cannot operate without and attackers who are ceaselessly finding vulnerabilities in them. For them, social engineering isn't a product problem or a dialectic debate. It is a clear and persistent problem—one that threatens the very survival of their organizations.

As I noted earlier, this frustration with the process of social science and the security vendor space led me to write this book. If you are reading this, you likely share many of these frustrations. You have realized that whatever has been done so far didn't help you pinpoint the weak links among your users, understand why they posed a risk, craft evidence-based risk mitigation solutions, or define your entire organization's social engineering vulnerability.

This book is the product of my principled frustration and the answer to yours. It distills years of research in cyber science, both my own and that of others, into actionable solutions in one comprehensive presentation. It is written such that these solutions can be directly applied without needing a vendor, a product license, or a subscription fee. For nothing more than your attention and willingness to try, you can use the knowledge and techniques from this book to solve the problem of social engineering.

### WHO IS THIS BOOK FOR?

If you are working to stop social engineering, this book is for you. If you are the IT manager creating policies to improve cyber resilience, the security staff member working to reduce the organization's exposure to social engineering, or the consultant helping them accomplish this, then this book is squarely for you. It is also of value to you if you work for the insurer indemnifying organizations, the law enforcement agencies working to protect them, and the regional and national-level policy makers enabling them.

The Cyber Risk Survey approach complements existing approaches for user training while significantly improving what's being done. The Cyber Hygiene Inventory provides the missing toolkit for quantifying and tracking users' thoughts and actions related to digital safety. Both the CRS and CHI fit readily with many organizations that already have a training pipeline that includes pen testing.

For such organizations, this book provides the missing framework for understanding and quantifying user risk. For organizations that don't already have a training protocol in place, it provides the framework for creating one, including the information necessary for developing a pen-testing approach from the ground up and for tracking cyber hygiene levels.

The book's contents are also of value to consultants and training organizations hoping to distinguish themselves in the marketplace. Using its approach, consultants can identify the users who are at risk and diagnose the reasons for it. Rather than prescribe more user training like every other consultant does, they can help organizations develop novel, evidence-based solutions that meet the needs of users.

This book is of value to cybersecurity insurers as well. By implementing the CRS across different organizations, insurers can assess cyber risk across a sector. They can do more than just audit the organizations presently at risk from social engineering. Now they can explain why their client organizations are at risk and provide them with ways to improve it. This helps insurers make more-reliable actuarial estimations and build better relationships with organizations they insure.

Finally, the contents of this book are also of value to national security and public policy makers looking to improve cyber resilience. By measuring cyber risk and cyber hygiene across a region, they can create applicable legislative policies. They can identify and target law enforcement and investigative resources to regions where they are most needed. These are more effective than the present vendor-provided estimates that have failed to deliver.

With the breadth of audience in mind, this book focuses on providing actionable knowledge. In place of lengthy commentary on social science theory and the intricacies of various malware, it uses case studies and vignettes. Instead of lengthy references and reams of empirical data, the book emphasizes brevity and explanation. The goal throughout is to move IT managers toward implementation of the CRS—so they can diagnose, detect, and defend their users before the next social engineering attack.

## HOW THIS BOOK IS ORGANIZED

This book will take you on a journey that leads to cyber resilience. The discovery begins in chapter 1 with an overview of how social engineering evolved into what it is today. We'll trace the root of this method of deception—the vector—to surprising places. We find it metastasizing in the highways of precolonial India. We'll discuss why this was the case and how, by applying the then budding science of criminology, a young British East India Company colonel ultimately defeated it. The lessons of that era, while lost to time, are even more relevant today—and we'll see why.

Chapter 2 examines the proliferation of email-based social engineering. Most people attribute the success of this vector of attack to email's ease of use and to the internet's ubiquity, but these are proximate explanations. They don't explain why other forms of online attacks aren't similarly successful. The chapter traces the ultimate reasons for the popularity of email-based social engineering to the supply and demand of user data, the fundamental flaws in the internet's system of authentication, the ease of finding users online, and the waning power of IT departments. These are system level causes that aren't going away any time soon. The chapter then explains how these have advantaged the social engineer.

Chapter 3 details the various approaches that organizations have taken to combat social engineering. It covers the different technical approaches as well as the more prevalent phishing penetration testing (or pen-test) approach to improving security awareness. If you are in IT, you are familiar with most of these, especially the technical approaches. Keeping with the spirit of the book, the chapter delves less into each technique of protection and focuses more broadly on each approach and its inherent pitfalls and strengths. The chapter culminates with an in-depth look at security awareness training and what we know about its effectiveness.

Chapter 4 provides the missing scientific framework—one that should have been the starting point to combat social engineering all along: an understanding of why people fall victim to it. The chapter explains how people think and act online and how elements in an attack are crafted to deceive them. The explanation of how people think centers around their cognitive processing styles and their beliefs about online risk. The explanation for how people act focuses on reaction, nonconscious habits, and their precursors. We'll discuss how these interact and explain *why* people get victimized.

Chapter 5 goes a step further. It presents the Suspicion, Cognition, Automaticity Model (SCAM) framework and discusses how the patterns of thinking and behaving online causally determine phishing susceptibility. The chapter explains *how* victimization occurs. Just as the depth and intensity of coughing serves as a marker of potential human illness, the chapter discusses a key symptom in the SCAM that acts as a marker of user susceptibility to phishing. This marker serves as a single convenient, quantifiable measure of deception detection.

Chapter 6 focuses on another missing part of the cyber risk assessment: a reliable pen test. The chapter explains how present-day pen testing follows no theory, guiding framework, or logic. Other than trying to trick people into clicking, there isn't much of a governing rationale for any test development. Because of this, much of the data from pen tests is not comparable and should not be used to diagnose risk. This chapter presents a solution to this problem: the Vishwas Triad (V-Triad)—an empirically derived framework that helps in designing a reliable and valid pen test. You'll also be introduced to the Cyber Risk Survey(CRS), an approach for measuring the validity of pen tests and user cyber risk.

Chapter 7 pulls the preceding chapters together. It shows how the CRS can be implemented by IT departments to conduct a comprehensive user cyber risk assessment. It details the steps to be taken by IT managers, from their assumptions about risk assessment to how they should create a suitable pen test, establish its baseline validity, deploy it, assess its outcomes, and measure its impact. The chapter explains how the data from the test can then be used to generate a net resilience score for the overall organization.

Chapter 8 tackles cyber hygiene—often seen in policy making and IT circles as a solution for reducing cyber risk from users. But cyber hygiene is an idea without guiding principles. There is no clarity on what it entails, what it doesn't, and when we know that someone has enough of it. We'll discuss the pitfalls of this. The chapter then delves into organizational design theory and presents five governing principles that help define cyber hygiene and shape our expectations from it. We'll then discuss a measurement tool, the 25-question Cyber Hygiene Inventory (CHI), which can be used to track user hygiene. The chapter will provide the entire tool and culminate with a discussion on the links between cyber hygiene and cyber risk, and how the CRS and CHI can be used in tandem to create cyber resilience.

Chapter 9 provides real-world examples of the cyber risk assessment process being implemented. It demonstrates how organizations of different sizes faced with a variety of challenges have implemented the CRS and CHI to assess cyber risk from social engineering, create a culture of security, and achieve user resilience against cyberattacks. This chapter provides working examples that elucidate how organizations have approached cyber risk from users and achieved resilience. At the end of this chapter, you will recognize the possibilities that lay ahead of you.

The book concludes in chapter 10 with a look into the future. It examines the changing IT landscape and discusses how our present security approach to users will likely make things worse for enterprise security. We'll look at how the advantages in favor of the hacker will tilt all the more so. We'll then assess how the CRS and CHI help redress this imbalance and reduce these advantages. The book's journey ends with a discussion of how doing what's advocated in the book will lead to a change in the role of IT from that of a first responder, showing up to put out fires, to that of an agent of change.

This book will change the way you diagnose, detect, and defend your users. It will revolutionize the way you combat social engineering. Let's begin the journey.

# 1 HOW SOCIAL ENGINEERING EVOLVED

If you traveled by road in India in the nineteenth century, chances are high that you died a terrible death, not by sickness or disease alone but likely at the hands of a Thuggee—a member of a vicious cult that plied the subcontinent and specialized in robbing caravans.

What connects the nineteenth-century Thuggees to twenty-first-century social engineers? To answer this, we need to travel back in time to the early 1800s. It was the time of the Industrial Revolution, when the roadways all across the Indian subcontinent were filled with caravans transporting raw materials from the colonial hinterlands to various ports and on to factories in England. Trading caravans served as the lifeline of all the chiefdoms, fiefdoms, princedoms, and kingdoms that had become vassals of the British. They transported people, products, and ideas—in essence, serving as the medieval information superhighway. Robbing caravans was good business, and the Thuggees excelled at highway robbery.

Their modus operandi was to win the confidence of travelers, isolate, overpower, and strangle them, hide their bodies, and sell the stolen merchandise using local support networks. Thuggees would spend months collecting information, impersonating travelers, winning their trust, and setting the trap. The cult developed its own language and codes and its members performed their ghastly activities under the auspices of a female goddess. By the 1800s, she had sanctioned the death of over a million travelers on the Indian subcontinent. Most simply disappeared while en route somewhere, never to be heard from again.

So choked was the overland movement of caravans that the British colonial administrators had to act. They outlawed thuggery and established the Thuggee and Dacoity Department, appointing William Henry Sleeman as its first superintendent. Sleeman, a young army captain with a flair for language and administration, developed intelligence-gathering techniques, novel for the time. He nurtured informants, charted family trees, and created cult member profiles, tracking their movement patterns, identifying burial grounds, and eventually deciphering their coded language.

Within a few years, he had captured their leader "Feringhea"—who led him to 1,400 other cult members who were subsequently incarcerated or hanged. In short order, the Thuggee cult that had brazenly operated for almost five centuries was completely eliminated.

Sleeman's success against Feringhea was heralded by the English press. It fed the imagination of British audiences, rich from their colonial conquests and from the steam-powered efficiencies of the Industrial Revolution. Sleeman's reports even inspired a best-selling ethnographic fiction, *Confessions of a Thug*, penned by Philip Meadows Taylor in 1839—which forever added the word *thug* to the English lexicon.

According to Joseph Hatfield of the US Naval Academy's Cyber Science program, it was around this time that the term *social engineering* first appeared.[1] British economist John Gray used it in his 1842 book *An Efficient Remedy for the Distress of Nations* to describe how engineers could tackle societal problems.[2] By using their advanced knowledge to study, influence, and inject corrective action, they could engineer solutions to all social ills, just as they fixed broken steam engines. It was the process Henry Sleeman had just used on the Thuggees, and it was how social engineering came to be defined for the next century, until another group of robbers during another revolution—this time led by telecommunications—would redefine it.

## FROM PHREAKERS TO HACKERS

In the 1950s, a 10-minute phone call from New York to Los Angeles cost $6.70, or $70.18 in today's dollars.

Not everyone was happy about this. It was the time of the AT&T monopoly, when the entire US telecom network was operated by the Bell System. There wasn't much that could be done, but one group did. They began robbing the companies. They called themselves "Phreakers." Their craft, "Phreaking," involved making unauthorized local and long-distance calls by circumventing phone charges. They were the progenitor of the modern-day computer hacker.

Most phreakers were students in engineering—a field that in the 1950s and 1960s was exploding with interest. Engineers, who accounted for roughly 3 percent of students in previous decades, now amounted to over 12 percent.[3] The field was also graduating twice as many students, most of whom planned to work for the booming telecom industry. Budding phreakers learned about the phone signaling system—how pulses, tones, and frequencies worked for making calls—in their classes.

They began by mimicking the signaling systems they had learned about in school using different approaches. For instance, one phreaker, Joe Engressia, was blind but could whistle and mimic the tones. Others experimented with different musical instruments.[4] Information about what worked was shared through all-night conference calls and newsletters such as TEL (Telephone Electronic Line). They were like graffiti artists, proudly showing off their skills. In time, this led to electronic devices of different types, dubbed "black, blue, and red boxes," where different tones for making free calls could be recorded and sold. They sold these to other phreakers and to consumers, motivating even more interest in phreaking.

At first, the phone company didn't care much. It tried to scare the phreakers by using the courts. When this failed, it developed internal fraud surveillance systems. One system, Project Greenstar, monitored 33 million phone calls between 1964 and 1970 and wiretapped over 2 million conversations—something we would consider illegal today.[5] But it helped them finally grasp the true extent of phreaking.

This started a cat-and-mouse game, with the phone companies (plural after the breakup of AT&T) developing technical countermeasures using newer frequencies, call trackers, and policing. In turn, phreakers turned into

Thuggees. They snuck into Bell South offices, wired up their own telephone lines, showed up at telephone company conferences dressed up as employees, and dug through company trash bins to find secret information. They discovered that certain frequencies, such as 2,600 Hz on US networks, could provide operator access to networks—analogous to getting administrator access to today's computer networks. When the phone company used digital filters and blocked these frequencies, phreakers developed war dialing techniques, creating automatic scanners that continuously probed networks all over the world, looking for open frequencies still susceptible to phreaking and then routing calls through them.

With the advent of mainframe computing in the 1980s, the phone networks began adopting cryptographic protocols and more-complex authentication methods. In response, phreakers began collaborating with students from the emerging field of computer science, who, like them, wanted to break into or hack systems. They chronicled their efforts in magazines such as *2600: The Hacker Quarterly*, its title reflecting the emerging bonds between phone phreakers and the new breed of hackers. It was within this group, in 1984, that the term *social engineering* reappeared—this time to describe the act of deceiving someone to make them unwittingly disclose technical information. It is also where social engineering acquired its darker meaning and entered the lexicon of hackers.

These collaborations continued until internet telephony eliminated long-distance call charges and made phreaking obsolete, but by then it had already influenced many young hackers, including Bill Gates, Steve Wozniak, and Steve Jobs. In time, as computing systems became more complex, hackers reverted to the techniques they learned from phreakers—social engineering to procure unauthorized access to computer systems.

While tracking the evolution of social engineering, Hatfield found two emergent perspectives. One, inspired by economists such as John Gray, focused on centralized planning by governments, political parties, and government departments to manage social change and influence the actions of people in a society. The second emerged out of phone phreaking. It implied the use of deception to gain access to computer systems and networks.

The first view led to streams of research in economics, sociology, anthropology, and political science, eventually falling out of favor in the 1990s, after the Soviets demonstrated the colossal failure of macro-level social engineering. But its principles are still very much alive. They form the basis of many public health interventions, from those that aim to make people stop smoking to those advertising the need to exercise more. It is also what today's IT managers are attempting with their cybersecurity awareness campaigns.

However, it is the second, negative view of social engineering that presently dominates. From reports in the media to policy discussions all over the world, the term *social engineering* today connotes online deception, usually some form of email or social media–based attack on users.

According to Hatfield, both the positive and negative conceptualizations of social engineering have much in common. Both involve an individual or group with an *information advantage* and a *knowledge advantage* over another. Information advantage comes from having access to better data and insights, knowledge advantage from having superior skills and technical know-how. Together, they allow the advantaged group to replace the goal of the other group with their own.

The phreakers, for instance, originally learned their skills from engineering programs they were enrolled in, and they shared information with each other through their all-night conference calls and newsletters. This gave them their information and knowledge advantage. The thrill of being the first to overcome a technical problem motivated phreakers to collaborate with outside groups such as hackers and share these novel solutions—which further inspired others in the group to do likewise and led to an upward cascade in information sharing and know-how.

But, unarguably, the telephone company had more highly trained staff and deeper pockets, with which they could have acquired information and stopped the phreakers sooner. So why didn't they? The answer can be found in two additional factors: the *incentive framework* and the *information environment* that existed in the telephone arena.

Until the antitrust breakup of AT&T in 1984, the telephone system in the US was a monopoly. It was a vast company with no competition and huge revenues from enormous markups—a mind-set that changed little even after the breakup. As Joseph Nacchio, the president and CEO of Qwest Communications Corporation, said in a 1997 *Business Week* interview: "Long distance is still the most profitable business in America, next to importing illegal cocaine."[6]

So, there was no incentive to stop the phreakers. The phone companies just didn't believe phreakers could make a dent in their revenues. It wasn't until the phreakers were getting attention in the media that the phone companies even began commissioning research studies to understand the scale of the problem.

The information environment was also transparent and accessible. Phil Lapsley,[7] who has researched the history of the phreakers, found that back in the 1960s most of what organizations today would label as intellectual property and corporate secrets was openly published in AT&T's technical journals. It was as if AT&T was showing off—and why wouldn't it? The last thing it was worried about was someone using its technical information to break into the only business model that existed: its own. Thus, the telephone company's sense of invulnerability furthered the information and knowledge advantages of the phreakers. This allowed the phreakers to fester and, in time, improve their skills and thrive.

Such factors, however, are not limited just to the telecom space. They are common in all forms of social engineering and dictate whether the deceptive or the corrective version succeeds. The Thuggees, for instance, used impersonation techniques to gather intelligence about the caravans they planned to rob. Their information advantage came from being organized as a cult, with secret rituals, coded language, and signals. This allowed them to share information across vast swaths of the Indian subcontinent and avoid being detected. Many were also disaffected soldiers from local armies. Like the soldiers of the disbanded Iraqi Army who formed the Islamic State's (ISIS) fighters after the 2003 US invasion,[8] many Thuggees learned their deadly skills through military training, which gave them their knowledge advantage.

In contrast, the caravans, being always on the move, had little advance information about what lay beyond the borders of individual friendly kingdoms. Most wayfarers had a short-term goal. It was to get to their predefined destination, at which point their interest in the caravan ended. They never cared about the longer-term safety of the caravan.

Further disadvantaging caravans was the lack of armed guards and knowledgeable experts who could spot a thug in their midst. This was because the kingdoms along the way, which were perennially in conflict with each other, gave the caravans special passage. The one caveat was that they carry minimal weaponry. This was to ensure they weren't concealing a neighboring army in disguise, a hard lesson learned since the Trojan War. This skewed the information and knowledge advantage against the caravans, making them easy targets.

Adding to all this was the caravans' business model. Much like today's social media platforms, their business model was one of scale. Larger caravans attracted more travelers, who sought safety in their numbers. This meant caravans kept adding people—a critical aspect of the incentive framework—making it easier for Thuggees to join and then take advantage of them.

Finally, during colonial times, the Indian subcontinent had close to two hundred different languages and some 550 dialects. Caravans would transport people from different ethnolinguistic groups and pass through kingdoms governed by rulers from various groups. The lack of a unifying language and the highly fractionalized administrative structure made record keeping hard. Most princely states kept records for their protection from neighboring kingdoms—which before modern transportation was where all conflicts occurred. No one really cared what happened many kingdoms down the caravan route. In fact, some rulers even employed Thuggees as mercenaries against their neighbors, which reduced the impetus for ratting them out. They willfully kept the information environment closed, contained, and proprietary. This was why, even though there had been rumors of Thuggees for centuries, their presence wasn't confirmed until the British came into India.

Henry Sleeman's quick success over the Thuggees came from an inversion of these very factors. Colonial officers were educated in military academies, such as the East India Company Military Seminary at Addiscombe, that distilled the varied experiences of the British Army from its colonies all over the world. Because of this, Sleeman was versed in engineering, linguistics, and surveying techniques. This knowledge advantage allowed him to design his novel criminal profiling skills and check the information advantage the

Thuggees enjoyed. The English language, which all colonists shared, unified the collection and sharing of data. The language was unknown to the Thuggees, as were the refined data-gathering techniques that the English education system had taught Sleeman. This gave him the information edge.

The British colonists also administered centrally. This gave Sleeman the information environment necessary for tracking Thuggee movements across the subcontinent and cracking their codes. His final advantage was the incentive framework. While each princely state was motivated to maintain the integrity of its borders, the British in India valued the trade routes across the country. Until the development of railroads, this was the primary mechanism for moving goods from the hinterlands of the colonies back to England. In other words, the safety of the caravan's entire route from start to finish, not just the stops along the way, was essential to their business model. Ultimately, the positive knowledge and information asymmetries within a motivated and supportive information environment allowed the British colonists to outengineer the Thuggees in India.

## THE EVOLUTION OF PHISHING

Monkeys are hard to catch. They are quicker than we are, nimbler, and have opposing thumbs, making them difficult to ensnare using most animal traps. In the south of India, villagers use an ingenious trick to catch them. They chain coconuts with sticky, sweetened rice onto the trees near where the monkeys are perched and wait for a curious monkey to be enticed by them. The coconuts are hollowed out such that the opening is just small enough for the monkey's palm to enter and grab the rice but too small for it to exit with its fist clenched full of rice. Once a monkey reaches for the irresistible treat, it gets so committed to the task that it forgets to unclench its fist. The monkey ends up trapped because of a sunk-cost fallacy, where the thought of recovering the tasty morsel for which it has already committed itself is all that it focuses on. The South Indian Monkey Trap—popularized in Robert Pirsig's philosophical novel *Zen and the Art of Motorcycle Maintenance*—works not by physically trapping the monkey but by playing on its habits of mind.[9]

Although we humans are much more evolved, we remain susceptible to similar mental traps, and some scammers have developed ingenious ways to use information and knowledge asymmetries to create such traps. While cults like the Thuggees took direct advantage of wars, upheavals, and revolutions, these scammers took indirect advantage of them.

Until the modern nation, boundaries between kingdoms, and with them the fortunes of people living in them, were constantly in flux. There was always someone in some nearby land falling out of someone's favor as the newer, more powerful overlords came into power. Before global news services, people usually heard about such events through word of mouth, town criers, or the mail, but there was no objective way to confirm the news. This gap in information and knowledge was fodder for opportunists. They developed scams around these events and took advantage of the burgeoning postal services to execute them.

Their con letter went as follows: "The letter . . . is written as fairly well-educated foreigners write English, with a word misspelled here and there, and an occasional foreign idiom. The writer is always in jail because of some political offense. He always has some large sum of money hid, and is invariably anxious that it should be recovered. . . . He is willing to give one-third of the concealed fortune to the man who will recover it." While you might recognize this as the Nigerian phishing email, this was a *New York Times* report from 1896 describing scam letters targeting Americans during the Spanish-American war.[10]

These scams worked because people were beguiled by the promise of riches and the small "advance fee" required to procure them. Once victims provided the advance, they were ensnared by the same process that trapped the curious monkey: they became unwilling to unclench their proverbial fist in the trap. The size of the potential prize and the thought of recovering the fee already paid—a sunk cost—becomes their primary mental focus, for which they are willing to pay some more in fees. The scammers, for their part, keep delaying and asking for a bit more money from the victims, who keep escalating their commitment and over time spend a lot more in trying to recover the fee they had advanced.

Finn Brunton, who researched the history of advance fee frauds, found reports of them from the time of the French Revolution.[11] An intriguing figure from this era, French criminal turned detective Eugene Vidocq, described its use and relative success. "Of a hundred such letters," he contended, "twenty were always answered."[12] It was an odds gambit whose success rate would rise as postal and banking services evolved and the scam could target people farther away.

From France, advance fee scams made their way through different war-torn regions. They showed up during the Spanish-American War, the Russian Revolution, and the world wars. Most still targeted people living geographically close to the warring states. With the development of telegraphy and radio, this changed. Scams began exploiting people in faraway lands, wherever the news might resonate or evoke response. For instance, in 1922, capitalizing on news of the postwar reconstruction efforts in Germany, Americans were targeted with circulars enticing them to pay a small advance fee to receive home beer-brewing kits from failing distillers in need of investments in Germany. The scale of all such scams increased even further with international postal mail and global banking. By the 1980s, global news networks such as CNN added more scope to the scams by presenting news of wars and corrupt governments in different parts of the globe. Now, more people in the world were familiar with such events.

Taking particular advantage of this were scammers in Nigeria, who, driven by the nation's poverty and systemic corruption, launched one of the most prolific versions.[13] Often called the Nigerian scam, letters claiming to be from a deposed minister with millions of dollars of illegal kickbacks from the country's recent petroleum boom in some slush fund were being mailed all over the world using counterfeit postage stamps. The story line was like the older versions of the scam: provide a local bank account and pay a small fee for clearing the funds in return for a percentage of the megamillions. And just like the earlier scams, this, too, was an odds gambit, with scammers sending millions of letters worldwide, hoping to entice a few responses.

The difference was its scale. Over a three-month period in 1998, US postal inspectors intercepted 2.5 million Nigerian scam letters at New York's John F. Kennedy airport; inspectors in the UK intercepted 1.5 million. In

2009, US postal inspectors intercepted close to half a million dollars in counterfeit postal money orders from Nigeria—all in one package.[14] The sheer scale made it likely that even a small percentage of victims sending in their initial advance fee could net the scammer enormous dividends— payoffs that could go up substantially as those victims attempted to recoup their sunk costs. This was big money for many poor nationals of Nigeria, where over 70 percent of the population lived on less than $2 per day.

Many Nigerian scammers prided themselves on their skills and looked at scamming as a craft—which often required flying their victims to nearby Benin, Togo, and Ghana, wining and dining them in fancy hotels, and having them meet fake business partners. This created supporting businesses, from fake printing presses to bankers willing to route foreign currency illegally. All this success attracted even more people in surrounding countries to try their hand at the scam.

By attracting more scammers, the scam became a victim of its own success. What until the 1960s had been the advance fee fraud now became known as the Nigerian 419 scam, a name derived from the designation in the Nigerian criminal code that outlawed it. The increased activity also drew the attention of international police and the media, who would forewarn unsuspecting victims, stopping them before they cognitively committed to the process. Many post offices instituted awareness campaigns, while banks trained their staff to stop people before they remitted money to countries like Nigeria. This diminished the information asymmetry the scam enjoyed. To survive, the scam needed victims who weren't knowledgeable about it and could become privately committed to the process before someone saw them or talked them out of it. This was made possible by the next technological revolution—the internet.

Starting in 1999, as internet cafes began proliferating throughout Nigeria, scammers took their 419 mail scams online. It turned into what has become the infamous Nigerian email scam, some form of which, thanks to the ubiquity of the internet, has been received by anyone who has an email account no matter where they are in the world. Free email, anonymous communication, and the ubiquity of email truly enhanced the reach of the attack and its scope. Further loading the deck against victims was the ability to remit money online, the lack of globally applicable laws on the internet, and the private consumption of email. This made it possible for victims to send money quickly. They couldn't be talked out of it at the bank or post office, and there was little legal recourse after the fact. Now the odds were even more in favor of the scammers.

As the scam evolved, so did its story line, which now incorporated Libyan generals, Eastern European oligarchs, and Middle Eastern princes, promising riches, brides, lottery winnings, cash, and cars. It also began attracting new perpetrators—Asians, Russians, Europeans, and even Americans—each learning from news reports of others. It is estimated that by 2005 there were over 250,000 scammers perpetrating advance fee scams worldwide.[15]

The attacks developed further as internet telephony made it possible to call people across the world for free. This attracted fresh actors with newer skills to the stage. In July 2018, the US Department of Justice indicted 21 individuals running a massive telephone call center network from India that targeted Americans by claiming to be the Internal Revenue Service (IRS).

Their modus operandi was the same as with earlier frauds. Using threats and warnings, they coerced people to make a small upfront payment to write off the rest of the sizable back taxes they alleged were owed.[16] The entire operation was professionally managed. It employed over 350 callers, with trained managers and supervisors who ran the scam like a legitimate, for-profit corporation would. Teams of analysts scoured the internet for information about their targets from other breaches. The scammers therefore had asymmetric information and knowledge advantages over their targets—which they leveraged. This made them enormously successful. Over a five-year period, the fraud netted the center close to 150 million dollars, with one victim alone losing $150,000.

According to the FBI's Internet Crime Complaint Center (IC3), in 2017 Americans lost close to half a billion dollars through online advance fee scams.[17] Other nations with poorer tracking, weaker enforcement, less-knowledgeable citizens, and fewer avenues for helping victims are ostensibly suffering even more. Global losses from Nigerian-type scams are estimated at over $12 billion annually, with billions more likely going unreported because victims are often too scarred or too embarrassed to report their losses.

Among the more publicized reports are those of a Japanese businessman who was lured to South Africa, kidnapped, and held for ransom and of a Czech retiree who, having lost $600,000 to the scam, ended up killing an official at the local Nigerian embassy.[18] The Thuggees, phone phreakers, and advance fee frauds had truly come of age on the internet.

Paralleling the Nigerian scam was another type of attack. It was far simpler but more insidious. It began unremarkably around 1996 on America Online (AOL), which at the time was the world's biggest internet service provider (ISP), with five million users. The attackers focused on stealing users' passwords through AOL's popular instant messaging service by posing as AOL employees and requesting that users verify their login and account password—their credentials.[19] The attack was using many of the social engineering tactics developed by Thuggees and phreakers. It employed impersonation, used a trustworthy name as bait, and targeted users directly. The new messaging platform allowed the perpetrators to stay anonymous and avoid getting caught. Because the platform was free to use, they could cast as wide a net as possible, trying all kinds of requests for information, repeating and perfecting them until they could net a victim. They could harvest the victim's credentials to create new AOL profiles and use them to entrap still more users. The scam became known as *phishing*— indicating the attack's roots in phone phreaking and its similarity to fishing.

Even AOL, where every connection to the internet went through a centralized service infrastructure, couldn't stop the *phishers*, the term given to the perpetrators. The largest ISP in the nation resorted to asking users to be vigilant by placing warnings on emails and instant messaging clients. The leading internet company of the time was behaving like the post offices and banks trying to stop the Nigerian scam's victims. The AOL community is where phishing first emerged, and it is also where we see organizations focusing on improving user awareness—a paradigm that, just like phishing, continues to this day.

By 2004, thanks to the explosive growth of internet technologies—high-bandwidth connections, free email services, and the rise of online shopping websites—phishers took to credential stealing on new and popular email services such as Yahoo and Hotmail. The old AOL attacks were rapidly evolving. Attackers were now crafting fake or spoofed websites of email login pages and re-creating online shopping portals, banking websites, and payment portals to steal users' credit cards and identities.

Scammers old and new were attracted to this attack vector. Some advance fee scammers in Nigeria began organizing into phishing gangs. One such gang was the Black Axe Syndicate—a notorious fraternity of criminals, much like the Thuggees, who specialized in the creation of highly credible, targeted phishing campaigns.[20] Working alongside them were different computer programmers, who until recently had been creating disruptive viruses but were losing their dominance because of the success of antivirus programs. For them, the email-based attack vector was attractive because it allowed them to focus on their craft of building malware while letting others, like Black Axe, use it. This turned virus makers into specialists creating "malware-for-hire."

The combination of low-tech phishing with high-tech malware-for-hire made for a potent combination. It brought together the devious minds working to scam people with the technically minded ones who could develop sophisticated exploits to enable them. This attracted attackers of all stripes, from petty thieves and organized criminal gangs to disgruntled employees, online activists ("hacktivists"), terrorist groups, and eventually government espionage agencies and even law enforcement.

An indicative example is BellTroX InfoTech Services, a little-known Indian IT firm that had provided hacking services to entities worldwide. According to a 2020 Reuters investigation, over a period of seven years, this firm had spied on more than 10,000 people, from judges in South Africa to lawyers in France and environmental groups in the US. They accomplished all this by bombarding their targets with spear phishing emails. Attacks would begin with broadly applicable emails (e.g., links to horoscopes), escalate to contain pornography, evolve into more-personalized emails (Facebook login requests and notifications to unsubscribe from pornography websites), and eventually graduate to imitating colleagues and relatives.[21]

There were many other firms, such as Israel-based NSO Group, which developed Pegasus, a highly potent malware that could be transferred onto phones via messaging apps, to others working in the dark corners of the web. All were primarily using some form of social engineering, each attracted by its simplicity: how little it took to craft a targeted attack using widely available online information, how simple it was to borrow technical skills to entrap users, and how easy it was to net victims. The proverbial monkey trap had evolved. Now its targets didn't even have to reach inside the trap—all they had to do was touch it, virtually, with a computer mouse or touch pad.

Soon, phishing traps were being found all over the internet. The FBI's 2017 IC3 tracked over a billion dollars in losses from phishing—amounts that had almost doubled from their previous year's report. By 2020, the losses had increased to $4.2 billion, and these were just immediate losses from different hacks. They excluded any losses to reputation, or ancillary costs such as the cost of intellectual property investments (e.g., Boeing spent $3.4 billion over 14 years in developing the C17 transport aircraft and had its plans stolen by the Chinese government), the loss of personnel (as we witnessed after the Sony Pictures breach), the loss of human life (as with the Ashley Madison email breach), or, as in the case of the Democratic National Committee, the ostensible loss of an election.

## CLASSIFYING PHISHING

Today's email-based attacks used by hackers targeting organizations can be broadly classified into two types. One involves some form of interactive, back-and-forth communication with the victim. These attacks, which are a derivative of the advance fee scam, use a ruse or pretext as the lure, so they are commonly referred to as *pretexting*. The second form of attack requires that a victim open an emailed attachment to deploy malware, visit a fake or spoofed page from which the malware is dropped, or enter their login and password credentials on a spoofed page. These attacks utilize different requests but require little to no communication exchanges with the sender. Such attacks are commonly termed *spear phishing*.

For consistency, this book uses these two terms throughout: *pretexting* when referring to any attacks requiring communication with a hacker and *spear phishing* for any attack deploying malware or a malicious hyperlink. The book uses the terms *phishing* or *social engineering* as umbrella concepts that subsume the two forms of attack and uses *phisher*, *hacker*, or *social engineer* for anyone who uses such attacks.

Of the two forms of attack, spear phishing is more virulent, because of the payload it could carry and how little someone has to do to fall victim to it. All the user needs to do is click on the malicious hyperlink or attachment. At the low end, malware hidden in attachments and on websites could be used for collecting intelligence that could subsequently lead to more-targeted attacks. At the high end, the malware could provide direct access to the user's device, from which the hacker could move laterally across a computer network, collect even more information, exfiltrate data, or even lock out every connected device in the network and hold the entire enterprise hostage in a "ransomware" attack.

For close to a decade now, spear phishing attacks have accomplished all this and much more. Chinese hackers used them to steal blueprints of fighter jets from Lockheed, Boeing, and other defense contractors and steal sensitive information from major US corporations such as the health insurer Anthem Inc., and from the US government's Office of Personnel Management. They were also used by Iranian operatives to hack into dam controllers in the New York area, by the Syrian Electronic Army for hacking into Twitter accounts of the Obama White House, and by Russian and other operatives for launching ransomware attacks on hospitals in California and New York. Global news media reports document the use of spear phishing by ISIS terrorists for locating and assassinating Syrian sympathizers and by North Korean hackers to target banks all over the world. It has also been used to attack industrial control systems in Germany, electric grids in Ukraine and Israel, oil and petroleum operations in Saudi Arabia, and political parties in Asia, Europe, and Latin America.

But while spear phishing is the favorite attack vector, pretexting attacks have also become more potent. The Nigerian email scams have evolved to compromise businesses by impersonating someone working for a company or by posing as an outside vendor. These attacks are more targeted, with the hacker learning about the organization and how it operates. In a telling example, Barbara Corcoran, the host of the popular television show *Shark Tank*, lost close to half a million dollars to a pretexting email that impersonated her assistant's email and sent a $400,000 invoice to her accountant. Since her assistant routinely sent such invoices mere seconds after receiving the email, the accountant wired the money.[22] The attacker knew how invoices were generated, how Corcoran's assistant sent them, and how Corcoran's accountant operated. In another attack, in 2016 a hacker pretended to be Snapchat's CEO, Evan Spiegel, and procured payroll data of all current and former employees.[23] In yet another attack, the US Department of Justice indicted a Lithuanian citizen for using fake vendor invoice emails and collecting $100 million worth of payments from Google and Facebook.[24]

Such attacks are even more successful when used in conjunction with spear phishing. Examples are the cyberattacks that targeted the Society for Worldwide Interbank Financial Telecommunication's (SWIFT) network, the proprietary communication system used by over 1,100 banks in 200 countries for executing interbank transactions. The hacks, attributed to the North Korean regime by the United Nations, stole over $2 billion from banks in various parts of the world.[25]

To accomplish this, the hackers first spear phished banks and installed backdoors for surreptitiously entering their computing networks. Next, spending upward of 155 days (and in one case more than 678 days) within each network, the hackers learned how each bank's individual staff members behaved. Using this knowledge, they used the SWIFT messaging system to generate emails asking the bank's staff to transfer sizable sums of money to other member banks. From there, the money was withdrawn, never to be seen again. The attacks were so well choreographed that the hackers even disabled printers in these organizations to eliminate paper trails.[26]

Today, phishing—spear phishing and pretexting—accounts for close to 90 percent of all social engineering attacks. *Verizon Data Breach Investigation Report* (DBIR), an annual study that tracks cyberattacks across two thousand organizations, found that one in three reported breaches in 2019 was because of phishing—a trend that has remained more or less the same since 2015.[27] Another tracking study, FireEye-Mandiant's 2018 *MTrends Report*,[28] reported that 50 percent of organizations targeted by social engineering seemed to be attacked again within a year, ostensibly with greater precision, as the hacker learned from the organization's reaction to the previous attack. Making things worse, the average global dwell time, or the time it takes for a successful attack to be discovered, rose to

186 days—another staggering trend that remains unchanged. Thus, social engineers continue to work with impunity, attack repeatedly, and remain concealed within most devices and networks for about six months before ever being discovered.

Standing in their path is today's IT manager. They are like Henry Sleeman, trying to protect the organization's users and defeat the virtual Thuggees. But, as we'll see in chapter 2, all the odds—the information advantage, the knowledge advantage, the information environment, and the incentive framework—are stacked against them.

# 2  WHAT MAKES SOCIAL ENGINEERING POSSIBLE

On March 19, 2016, John Podesta, chairman of Hillary Clinton's presidential campaign, received an email security alert from Google asking him to reset his Gmail password. After checking with his chief of staff, Podesta complied.[1]

The alert was part of a spear phishing campaign targeting Podesta's email account. Russian state-sponsored hackers used Podesta's account to download 20,000 campaign-related emails. Also targeted were staffers on the Democratic National Committee (DNC) and the Democratic Congressional Campaign Committee. Many complied, only to find their emails, like Podesta's, hacked and released through WikiLeaks and other file-sharing sites. The content of the emails, some taken out of context and others amplified through organized disinformation campaigns, eventually influenced many voters—and altered the course of US political history.

On a less history-making level, in September 2018, Apple Inc. reported that someone from Australia had hacked into its internal servers multiple times. This was after prior breaches in 2014, 2015, and 2017. The 2014 breach was the infamous iCloud photo leak[2] in which someone had illegally obtained and released compromising photographs of many celebrities. Given this, Apple was being extra cautious of data access on its network. Yet, it was breached again. This time it wasn't users' private photographs but instead 90 gigabytes of data stolen directly from Apple's corporate servers—a big blow for a company legendary for its culture of secrecy.[3]

The Clinton campaign breach and the Apple breach had two things in common. Both attacks were perpetrated using spear phishing and both victims also had advance knowledge of and experience dealing with attacks of this kind. In Apple's case, the earlier iCloud breach had resulted from spear phishing, while the FBI had warned the DNC they could be targeted.[4] The DNC also had an IT help desk with the technical skills to deal with such attacks, which Podesta's chief of staff had specifically asked about the email.[5]

So, what makes phishing so difficult to stop that neither advance warnings by the FBI nor the best technical minds can forestall it? Whenever this question is posed, the usual answer is that social engineering attacks are easy to craft. But that is a proximate cause. Sure, phishing, because it utilizes emails, phone calls, and text messages, is easy to craft, but so are other crimes. Breaking into and stealing from houses doesn't require much skill, because many people leave their front doors unlocked. Yet, such crimes are rare.[6]

Proximate causes often point to obvious influences, which can be satisfying because they fulfill our need for understanding. But they can obscure the ultimate causes of the problem, making us look for solutions in the wrong places. That is why people invest in home cameras and security systems to fight a smaller problem but are unwilling to spend money to protect their identity online, which is at a greater risk of being stolen. This myopic view of the causes of phishing is a reason phishing has been unstoppable.

This chapter explains the ultimate reasons phishing is hard to stop, which have to do with money, credentials, the size of the virtual world, and the "people problem" of cybersecurity.

### MONEY

Having prior knowledge and experience with spear phishing isn't the only commonality between the 2016 DNC hack and the 2018 Apple breach. There is another similarity: teenagers perpetrated both attacks. The hack into Podesta's email was by a 14-year-old Kazak-Canadian, Karim Baratov.[7] The hack into Apple was by a 16-year-old from Melbourne, Australia,[8] whose name remains sealed because of Australian laws protecting juvenile offenders.

Baratov had two skills. One was spear phishing—a practice he perfected after dropping out of school. Russian intelligence often sought his help to steal the email credentials of users, like Podesta, who didn't use Yahoo's email service, which they had already compromised through a previous password breach in 2013.[9] That's why many experts believe it was Baratov who helped hack Podesta.[10]

Baratov's other skill set was that he knew about American culture, customs, and holidays. With this, he could craft compelling phishing emails, which he did until the US Department of Justice (DoJ) indicted him at age 21.

Like Baratov, the teen who hacked Apple was a relative amateur. He used a consumer-grade laptop for his heist and kept the stolen data in a folder named "hacky hack hack." But they aren't the only teen phishers.

In July 2020, the Twitter accounts of Bill Gates, Elon Musk, Barack Obama, and many other well-known people in politics and technology were hacked. The hackers, who had tweeted out requests for a $1,000 donation from these accounts, included a 19-year-old from the UK, a 22-year-old from Florida, a 16-year-old from Massachusetts, and another minor from Florida, who had used phone-based spear phishing for their attack.[11] The 2015 and 2017 hacks into Apple mentioned earlier were also by a 17-year-old, from Adelaide, Australia.[12]

And take Marcus Hutchins, the 24-year-old credited with stopping WannaCry, the distributed ransomware attack that in May 2017 infected some 300,000 computers worldwide. Mere months after being feted in the media for this accomplishment, the DoJ indicted him for hosting phishing domains and creating password-stealing malware. Like Baratov, Hutchins had started his phishing practice in his teens and persisted with it until caught.

This hadn't been the case in the past. Early hackers did not lead to even more hackers. Many became professors, entrepreneurs, and technologists. This includes the likes of Robert T. Morris, who developed one of the first computer worms, the Morris worm,[13] and later became a faculty member at MIT. Bill Gates, Paul Allen, and Steve Jobs also engaged in phreaking and hacking in their teens, and we know how they ended up. So, there appears to be something about phishing that encourages even more of it. Understanding the reason for it unlocks one of the ultimate causes of phishing.

What encourages phishing is the demand and supply of user data. It is this market that fuels phishing—and makes it unstoppable. This only became possible when the computer went from being primarily a business machine to one used to manage all facets of people's lives.

Back in the 1960s, computers were extremely large and cumbersome, costing millions of dollars to purchase and maintain. Using these machines wasn't easy either. It required specialized coding skills for programming and use. Over time, advances in processing chips, file management systems, and eventually innovations in battery and display technologies led to the development of smaller personal computers (PCs).

The PC was a game changer because it cost less, stored files directly, and allowed the processing of data. This spurred the development of simplified methods for users to interact with computers, giving rise to graphical user interfaces (GUIs), standardized operating systems, and user-friendly software. The PC's ease of use attracted more users and with that more usage. Eventually, individual PCs and servers created the web of interconnections that became the internet of today.

Paralleling the growth in computing was the rise of hacking. Early hackers were technology enthusiasts who were showing off their technical prowess, not destroying the operating capabilities of PCs. Even if they desired to, most couldn't do much damage.

The lack of connectivity between computers curtailed this. Malicious viruses had to be physically ported between computers via storage devices. For instance, Brain, the first computer virus for MS-DOS, developed in 1986 by two Pakistan-based programmers, was distributed on floppy disks containing illegal copies of their software.[14] Likewise, the first ransomware attack, the AIDS Trojan, developed by a biologist in 1989, was ported on 20,000 floppy disks that he handed out during a World Health Organization conference on AIDS.[15] But this need for physical distribution changed when computers and servers became interconnected—when they became the internet.

The internet enabled viruses to be distributed widely through email programs. The first of these was the Morris worm, deployed in 1988 on ARPANET, the progenitor of the internet. This worm exploited weak passwords and other vulnerabilities in the UNIX mail system to find host machines, replicate, and spread. Estimates are that the Morris worm infected over six thousand major UNIX machines and hubs—about one-third of the extant network—and slowed down functions at many major universities, including Harvard, Stanford, and Princeton, and Lawrence Livermore National Laboratory.

Stopping it required partitioning of the entire network's traffic for several days at an overall estimated cost of $100,000—an inflation-adjusted value of around $220,000 in today's dollars. So, while the Morris worm was significant and became the blueprint for much of today's malware, its damage wasn't.

This was the case with many of the computer viruses that followed. They could destroy or disrupt but couldn't steal data. There were three reasons for this. First, because of the small amount of storage on individual PCs, there really wasn't much data available for stealing. For instance, a 1998 Apple iMac had a mere 4-gigabyte hard drive. Second, much of the data was also of limited value, because computing was still utilized largely by businesses and mainly for clerical tasks.

Third, the use of dial-up connections, where users connected to the internet via the telephone network, heavily curtailed the ability to steal. This technique of going online, which was used well into the early years of the new millennium, involved connecting to servers through internet service providers (ISPs), downloading data, and then disconnecting. The telephone companies billed on a per minute basis for internet connectivity, so users used the internet sparingly. They would disconnect from networks and shut down their systems. This curtailed both the spread of worms and, because the devices were no longer on the network, the ability to extract data from them. In short order, this changed.

Beginning in 2001, most developed nations started leveraging the existing cable television infrastructure and moved from dial-up to broadband. Users not only enjoyed higher internet connectivity speeds but were also no longer billed for the time they spent online. This prompted a massive increase in internet usage time and a commensurate increase in the amount of data being shared online. Users were also online more often, so their devices were rarely disconnected from the internet. With this came a new business model that had begun a few years earlier but would soon come to dominate the internet.

It started in 1996 with web-based email services such as Hotmail. At the time, ISPs like AOL provided most email services as part of their paid service. To increase signups, Hotmail offered free email in return for monetizing information users stored or revealed. Users loved the model. Within a year, 10 million subscribers had signed up for Hotmail.

Other businesses took note and followed suit. There were search engines, mapping services, messaging platforms, and, in time, social networking services, photo sharing, file sharing, and even entire operating systems. With this came more user data. This included data about users' communications, their private moments, and the inner workings of their lives, as email and messaging became more ubiquitous, as well as data about where they were, how they moved, who they met, and how they met. In time, with more businesses coming online, data about each person's health, ancestry, and even their DNA became accessible. During the early years of the twenty-first century, there was more data being provided by users to online services than companies knew what to do with. Data remained a commodity, and its potential was yet to be realized.

This changed with the rise of search engines like Google and social media companies like Facebook, which figured out how to harness user data. Google not only gathered data about users on its services but also analyzed it for patterns. This became the basis for targeting users with advertisements and promotions. Facebook collated data both from its subscribers about what they liked and disliked and from those they were connected to, making it possible to segment users based on their affinities.

Advertisers flocked to these online businesses because of the fine-grained user segmentation opportunities they provided. They could dissect the potential market for their products in innumerable ways and microtarget users. User responses to online promotions could also be tracked, allowing measurement of advertising expenses at a level that mass media (TV, radio, newspaper) advertising never permitted. This made user data the currency of the internet. It dictated the market valuations of many online businesses.

In 1997, Microsoft acquired Hotmail for $500 million in order to access its 8.5 million email users' data.[16] The ability to monetize user data drove Google's purchase of DoubleClick in 2007 for $3.1 billion, twice what it had paid for YouTube.[17] This also led to Acxiom, a database marketing company virtually unknown to most consumers, being valued at $3.5 billion in 2020.

Such valuations led to even bigger data-sopping operations. Facebook patents revealed plans to mine the microphones of devices to capture ambient sounds that revealed their users' media usage (e.g., television viewing) patterns. Companies like Google launched an alphabet soup of services from free email to mapping and photo sharing. Others, including Amazon, Microsoft, and Apple, also jumped in with a slew of products and services. This led to even more data about users being collected, mined, and transacted online.

By 2020, Facebook was valued at $700 billion for its two billion users, which translated to a back-of-the-envelope $350 per user. Google was valued $1 trillion for its four billion active users, around $250 per user. These valuations reflected the demand for user data in the online marketplace and the quality of information about the user these companies could supply. The money chasing user data pushed more organizations online, and with it came even more granular user data.

This data demand ultimately drove social engineering. Using the very same free email and messaging services that were built to capture user data, hackers could get access to devices and steal that data. They could use this access to get into organizational servers, and because many more things, from accounting to finance and communication, were now online, accessing organizations' servers provided an endless supply of data that hackers could harness, repurpose, or sell. So pivotal was data to the organization's survival that rather than risk the embarrassment of losing it, companies were even willing to pay a ransom for it.

Lubricating all this was a marketplace—called the Dark Web. Here, hackers could transact their stolen wares and make money from it. Malware, hacking skills, and user data could be purchased in this market—for a fraction of the price Facebook or Google might charge. Users' stolen social security numbers could be purchased for $4 a record, credit cards with up to $1,000 in balances for $10, and compromised bank accounts with $10,000 in balances for $25.[18] Further enabling these transactions were untraceable online payment gateways, digital currencies, and encrypted communication apps, along with little to no policing or legal oversight. Finally, there was patronage. Nations, espionage agencies, and even law enforcement in many nations sought the same user data for espionage and for attacks against dissidents and foes. They nurtured hackers and emboldened them.

Valuable data, a market for selling it, and deep-pocketed patrons willing to buy it have contributed to the persistence of phishing. These factors allowed Marcus Hutchins to develop password-stealing malware and profit from it on the Dark Web. They also allowed Karim Baratov to profit from selling his spear phishing skills to Russian intelligence. Baratov made his first million at 15. By the time he was indicted at 21, he'd owned multiple Rolex watches, Armani suits, a BMW, a Lamborghini, a Porsche, a Mercedes, an Aston Martin, and a house worth close to $1 million.[19] With so much money chasing data that is so easy to capture through phishing, is it a surprise that many of today's teens are taking to phishing rather than starting technology corporations?

### CREDENTIALS

There is another reason phishing is preferred: what can be done using it. Phishing provides the easiest, most direct means for stealing user credentials, as in, their logins and passwords.

Credentials are stolen because, once acquired, it is virtually impossible for any computing system to discern the identity of the hacker from that of its legitimate users. It is the starting point of any breach, and all that is necessary for procuring full access to all the data in a computer network. It was John Podesta's Gmail password that was stolen in 2016, and it was iCloud users' and Apple employees' credentials that were stolen by hackers in 2014, 2015, 2017, and 2018.

Our system of using credentials for identity management on computers began more than half a century ago and was flawed from its very inception–a flaw that was obvious from its earliest days but ignored. It therefore festered and is another reason for phishing's popularity among hackers.

The credentialing system of today was developed during the early days of mainframe computing, at MIT's Compatible Time-Sharing System (CTSS). This project invented multitasking on mainframe computers in the 1960s and paved the way for personal computing. Allowing multiple users necessitated access management so each user's files could be kept separate. The simple solution was to provide each user with a password that only they knew and entered along with a login. The computing system then matched the password to the login internally for authenticating the user and providing access.

This had an obvious flaw: it was impossible for the system to determine whether the person entering the login and password was its legitimate user.[20] This wasn't thought to be a problem, because the system was meant to be used by only a handful of trusted users, all of whom endeavored to keep the integrity of the mainframe computer. They reasoned that, after all, no one benefited from disrupting the system.

Of course, they were wrong. The limitation was exploited right away by another MIT graduate student, Allan Scherr. In the first historical case of credential hacking, Scherr, who wanted more time on CTSS, printed out the password of every user from the system's database, logged in using their credentials, and poached their allotted time. The system wasn't any wiser, nor was his original sin publicly confirmed until he admitted to it in the late 1990s. But everyone in the hacking community already knew of it. His credential stealing had inspired many copycats, the most famous being the story of a young Bill Gates and Paul Allen hacking into Computer Center Corporation's mainframe to gain free computing time.[21]

Not much was thought of such hacking attempts. Just as phone phreaking was viewed by the telephone company, the original credential compromises were seen as minor acts by curious teenagers eager to use the new computing technology. So, when caught hacking using stolen credentials, 15-year-old Bill Gates was merely banned from using the system for a year. Lack of access was considered punishment enough. Thus, no serious attempt was ever made to strengthen the system.

Because of this collective indifference toward credential hacking, in time the login and password became the de facto way to access all online services. Today, every computer-based network, from the electric grid to hospital systems, dam controllers, nuclear power plants, the global banking system, and even systems on the International Space Station (ISS), is protected primarily by passwords.[22] It is this critical piece of information that every hacker—even those who recently targeted the ISS—is ultimately after, because once the password is acquired, the hacker becomes the user and enjoys all the virtual privileges of the user on the computing system.

In recent years, the deluge of breaches because of credential compromises has spurred several efforts toward improving the system. These include knowledge-based authentication approaches, where users are asked to answer questions only they are likely to know the answers to (e.g., their mother's maiden name, the name of their first pet, and such); two-factor authentication (2FA), where a numerical password is sent via a text message to a phone in the user's possession, to be entered along with their login and password credentials; physical tokens, where a digital key is provided that needs to be attached in order to confirm the user's presence; and biometrics such as facial identification and thumbprints, which are entered as proof of the user's physical presence during authentication. Some nations, such as Estonia, have even introduced electronic identity cards with encrypted chips in them that have to be used to confirm one's identity when online.

These approaches add different layers of additional information to the original login and password combination, but they merely slow hackers rather than stopping them. They also give users a false sense of security that ends up making them more vulnerable rather than less.

Take 2FA, one of the most widely implemented solutions for securing credentials. Support for it is provided by every major online service from Amazon to Zoom. But 2FA doesn't solve the authentication problem. This is because the second numeric factor, called the authentication token, has to be housed on a computer somewhere. It, too, is protected by another set of credentials, another password combination. Each solution layer is hidden behind another layer that suffers from the same limitation it is meant to protect. The fundamental weakness of the primary system is just transferred to the next one, which, when hacked, unravels the entire system.

Cybersecurity firm Mandiant demonstrated this in a simulated attack on an organization that used 2FA to protect its workstation. As all organizations do, the system administrator secured the authentication tokens on a central server by using a login and password. Mandiant's team directly hacked that server. Then, working backward, they identified the workstations the passwords belonged to and gained full access to them.[23]

Other simulations have demonstrated techniques for circumventing different layers of 2FA by sending fake SMS requests directing users to provide their second-factor token; faking or spoofing the web page where users enter their second factor;[24] faking the user's phone number on which they receive the token; or directly stealing session cookies from browsers that keep users logged into various accounts in order to bypass the need for even entering credentials.[25]

But while simulations show what is theoretically possible, even less is actually needed to steal credentials. Take the case of Deloitte, one of the world's top cybersecurity consultancies, whose client emails were compromised in late 2017 by someone directly hacking into its global email server, where multi-factor authentication wasn't even enabled.[26] Consider also the more recent hack into millions of Microsoft's Outlook, MSN, and Hotmail emails, where hackers stole the credentials of Microsoft's customer support agents. The agents weren't using two-factor authentication but because of their tasks had privileged remote access to client machines.[27] This meant that hackers had backdoor access into all customer devices through the remote portal. It didn't matter what password or layers of authentication the customers had implemented. The 2FA process merely gave them a false sense of protection.

Outside 2FA, many security experts advocate biometrics as a mechanism for secure authentication. Some futurists have even suggested implanting biometric chips and using DNA markers to authenticate users. On the surface, such solutions appear secure, but not all biometrics are unique or will remain so as medical science advances. Even human DNA, long considered unique to each person, can now be changed. This became apparent from a recent case where a bone marrow recipient's DNA changed into that of his donor. The recipient's lips and cheeks still had his own DNA, while his semen had the donor's.[28]

Biometrics such as facial features and fingerprints suffer from another critical weakness. If stolen, a biometric marker cannot be revoked or changed. Thus, once lost, the person suffering the compromise can never recover from that loss. Besides, stealing a biometric marker used for authentication isn't that difficult. They are, after all, digitized and housed on a computing device that is also protected by a password and login combination. Just like authentication tokens, the very weakness in credentials it is meant to overcome is used to secure it.

Thus, no matter how secure a system or the number of layers of protection it has, hacking it is just a matter of finding the credentials of an authorized user—one who has access to data of value. This is what spear phishing and pretexting help achieve.

## IT'S A SMALL VIRTUAL WORLD

Finding an authorized user with privileged access is far easier than most people realize. Before the internet, people were separated by about "six degrees of separation," where connecting with a complete stranger required connecting with at least five other individuals who served as intermediaries. In the hyperconnected world we live in, most of us are separated by just a click on the right search portal. This one click separates us from each other. It is also what separates hackers from privileged users.

The six degrees of separation concept was popularized by a parlor game involving actor Kevin Bacon. Players had to find the fewest number of others in their circle of acquaintances separating them from him. Google even developed a search algorithm that provides a Bacon number showing any Hollywood actor's degrees of separation from Kevin Bacon.

The basis of the game was a seminal 1960s research study by Harvard professor Stanley Milgram. Milgram randomly chose volunteers from Wichita and Omaha and asked them to transmit a folder by hand to a target individual in Boston.[29] The expectation was that the folder's transmission would vary based on the person's social connectedness, their status, and where they lived (the relative population density, geographical sparseness, postal system, etc.).

Surprisingly, none of this mattered. Regardless of differences in status and location, it took each volunteer five mutually connected individuals to reach their targets—leading to the conclusion that most people were separated by six degrees. Milgram called it the "small world" phenomenon.

But this was a different era. Not only were there no regional, national, or global repositories of information, even interstate telephone calls, as we saw in chapter 1, were prohibitively expensive. People had to travel to different places and ask well-connected locals, such as mail carriers, priests, bartenders, or clothing merchants. This took time, effort, money, and motivation.

The world is even smaller now. Email, social media, VoIP (voice over internet protocol), and intermediaries such as search engines, web pages, and blogs help locate and connect with anyone anywhere in mere seconds and at no cost. There is no need to interface with anyone or spend much effort going places. You can find out about people merely by moving a finger.

Data from replications of the Milgram study present evidence of this. Research examining interconnections between Facebook users finds everyone is separated by four others; similar research on Twitter finds evidence of three degrees of separation. Keep in mind that these are only the connection possibilities within each platform. They don't factor in cross-platform searches, such as when someone looks up something on a search engine or a public database and follows it up on different social media platforms.

But hackers don't even need to go through that much effort. In 2020, Zhenhua, a cybersecurity firm operating in China, was found to have methodically collated data from across the internet about politicians, journalists, dissidents, and technologists in the US, UK, Australia, and India. They had crawled every available online repository, used data from social media posts, and also added hacked data from the Dark Web. The scale of the database was staggering. Not only was there extensive data about the target individuals, but there was also similar data about their family members, relatives, and acquaintances. With a single click, Zhenhua's clients could locate people in power or those closely connected to them. They could find private information about them or anyone else connected to them. From there on, the limits on what could be accomplished depended on the motivation of the client.

In the online world of today, it doesn't take five steps for a hacker to find someone. It just takes one on the right database. We are separated from a hacker by one degree. This ease of locating and learning about people virtually and accessing authorized users is another reason phishing thrives.

### THE "PEOPLE PROBLEM"

"I hate users!" "Give me a network that doesn't have users and I will guarantee you security." These were the words of a chief information security officer

(CISO) of a US government research laboratory that conducts top-secret research expressing his fears about enterprise security during a meeting I attended. Such sentiments are rather common in security circles and reflect both the problems posed by users in today's organizational computing space and the frustrations CISOs face while dealing with them. But how did it come to this? And why were IT managers so frustrated with users? The answer reveals another ultimate reason phishing is so successful.

In the early days of mainframe computers, IT managers controlled access to everything people did on them. Back in those days, hardware and software failures constantly cropped up, and solutions had to be ingeniously engineered. Most early IT managers were therefore engineers with knowledge of electronics and software coding skills. They were also highly respected. In some ways, they were like nineteenth-century locomotive engine drivers, highly respected for what they could do with the enormous machines, which with computers meant taking in code and churning out information.

As computing changed, so did the role of IT managers. During the 1990s era of multiuser computing and Windows NT, they became network managers and engineers. With the World Wide Web and the rise of search engines, their primary role became that of information management, which was reflected in newer organizational titles such as database manager and chief information officer. In the decade since the SPE hack, IT managers' roles and titles have further evolved. Today they are CISOs and security officers, primarily responsible for information security within organizations. But while the roles and responsibilities of IT managers have increased, their control over organizational use of technology and their overall value within the organization have actually devolved—and it all started with the computer mouse.

The 1990s saw the development of graphical user interfaces (GUIs), where computer programmers developed action buttons that could activate different functions in software. All users had to do was use the computer's mouse to point and click on the button. This simplified word processing, database management, and presentation functionalities. Now users could do complex things on their computers without needing to know a single line of code.

In time, users could create email accounts, develop web pages, create computer networks, and maintain personal servers, using easy-to-use, off-the-shelf, third-party software and services, without ever needing to interface with an IT manager. The IT department was increasingly becoming a support function for users, only called when things went wrong with computing technology, such as when there was a virus infection or some other serious hardware or software failure that the user couldn't handle. Just as locomotive engineers' value diminished with the evolution of newer forms of transportation, IT managers' value within organizations had also diminished.

By 2006, with high-speed internet connections and the availability of increased bandwidth, this trend had worsened. Amazon, and then in 2008 Google and other vendors, launched server farms, offering external data warehousing, hosting, and IT management services. These helped outsource many in-house IT operations and reduce costs—a need that was felt even more strongly during the financial crisis that followed. The recessionary forces combined with the growth of vendors, many in nations with low labor costs, such as India and the Philippines, led to the outsourcing of many more in-house IT functions.

Even as the outsourcing trend was picking up pace, another major technological innovation came along. This time it was from Apple: the iPhone. Launched in 2007, the iPhone and its mobile application (or "app") market shifted the targeting of products and programs to the average user—who was expected to have minimal coding skills and be in need of a reliable and easy-to-use computing device. At the time, most PCs, and even smartphones such as Blackberry, were being sold to businesses, but Apple focused on selling a product that simply worked out of the box without needing much support.

Users could purchase their iPhones from the Apple store and also download their software directly from its app market. In time, many major technology companies followed suit, offering their operating systems, hardware, and software through their own marketplaces. This was a significant change in IT practice. Not long ago, new operating systems and software were tested within organizations by internal IT quality control processes, usually for at least a year, before being deployed and installed on employee workstations. This process made it easier to identify fundamental flaws in software and avoid flaws in the system unknown even to the software's creator—zero-day exploits—before they became widespread.

In the new paradigm Apple inspired, technology vendors and developers dictated what devices were purchased, what was available for download, and when and how it could be used. System updates and even critical flaws were soon being fixed through downloadable software patches released on vendor websites. Tasks that IT departments used to ensure, such as updating and applying critical patches, now became the users' responsibility. With most hardware, software, and troubleshooting functions managed directly by users or through vendor help desks, IT staff within organizations were no longer seen as necessary.

The ensuing years saw even more innovations in the portable computing space, with tablets, smartwatches, wearables, and internet of things (IoT) gadgets all being marketed directly to users. Users were lining up to purchase these products, at times before they were even officially ready for launch, through sites such as Kickstarter and GoFundMe.

The media spurred this further. They heightened the prominence of new technology by reporting on product launches, offering reviews, and unboxing videos. New technology developers and users were described as innovators and mavens compared to those eschewing technology, who were termed laggards, diehards, and Luddites. So, adopting the newest and most novel technology became the norm—which with the availability of more technological products became an insidious problem for IT managers.

Many users were adopting new, untested technologies without considering their security implications. They were performing work-related tasks on these devices and, worst of all, bringing them into the workplace. On the one hand, this trend gave rise to new organizational practices, such as telecommuting and BYOD (bring your own device) to work, which reduced the capital costs for businesses, but, for organizational IT, users became the largest source of connections to and from the organizational networks and their single biggest worry.

Their worries weren't unfounded. Many users were unaware of how their technology use practices could affect the organization. The CISO of the federal research facility we met at the beginning of this section related the following story. Because the facility conducts top-secret research, his users can't so much as listen to music, let alone visit any external website, check email, or bring a phone to work. Staff frequently complained about the poor quality of work life when compared to that of any of their counterparts in civilian organizations. To improve morale, the CISO finally decided to allow the use of personal iPods. In a subsequent audit, he found users charging these devices on the USB ports of their secure computers. Consider the security implication of this.

Besides such inadvertent actions, users were also knowingly circumventing policy. Some did it because it was expedient and security got in the way of their work efficiency. Others did it simply because it was convenient. For instance, during another meeting I attended, an employee of a leading security agency admitted to not applying updates on his official, government-issued smartphone. Why? Because the device also served as an entertainment device for his young child, and the patch would have blocked those applications.

Finally, there were security and privacy issues with many new technologies that were completely unforeseeable, even to IT managers. The US Army learned this the hard way when its personnel were allowed to use the health app Strava for tracking their daily workouts. The app posted users' workout locations online, which publicly exposed the locations of the army's secret military bases all over the world.[30]

Thus, in the years leading up to the Sony Pictures hack of 2014, the problem in the hands of IT staff was not technology. Those problems were patchable, protectable, and outsourceable. It was the technology user. For IT managers, users—people using technology who weren't complying with IT policy or who were flagrantly circumventing it—became the single biggest concern.

And this was a group that IT managers were least equipped to deal with. Since the mainframe days, most IT staff had been recruited from computer science and engineering programs, where the training on the social science of security was limited.

With IT ranks already depleted by outsourcing and layoffs, and lacking the skill depth to deal with users, IT managers were feeling squeezed.

On the one side, there was pressure from top executives and a rising crop of insurers who wanted guarantees of cybersecurity, but on the other, there were users who were unpredictable, uncontrollable, and irreverent. IT managers came up with a term to describe this predicament: the "people problem of cyber security."

The label fit, and it helped IT workers explain their woes, but because they were blaming users rather than trying to understand them, the user problem of cybersecurity worsened. It grew and morphed into even bigger problems, leading to the Sony breach and all those that followed. It is the gap in IT departments' understanding of users—which exists to this day— that phishers exploit. This is the final reason why phishing thrives.

## ADVANTAGES OF THE SOCIAL ENGINEER

In chapter 1, we discussed four factors—information advantages, knowledge advantages, the incentive framework, and the information environment— that determine the success of social engineering. Today, the quartet of forces stemming from the internet's commercial model, which monetizes users' data, its weak credentialing system, the ease of locating people with access online, and the "people problem" of cybersecurity have shifted the balance of power in favor of phishers.

### Information Advantage

Today, consumers leave trails of personal data on social media, search engines, email services, and mobile apps. Some data they willingly provide. Others, because of the lack of globally enforced laws against user tracking, are gathered without their knowledge by online data aggregators in different parts of the world (such as Zhenhua in China). Adding to this is data from over a decade's worth of breaches—health records, logins and passwords, answers to knowledge-based authentication questions, private photographs, travel itineraries, and banking and financial information—available in Dark Web forums. Much of this information cannot be repudiated, revoked, or replaced, which means anyone with access to that information can reuse it in perpetuity.

But it gets worse. Because of varying laws governing how organizations must notify users in the event of a breach of their data, users are often oblivious of the information about them that has been compromised. This information asymmetry makes it far easier for hackers to reuse stolen information and achieve a secondary compromise. This was demonstrated by a spate of spear phishing attacks in 2018 that reproduced stolen passwords and knowledge-based authentication responses from a prior breach. The hacker presented users with their old passwords as proof that their account had been compromised. Because users weren't aware of the earlier breach, most ended up entering their passwords and getting revictimized.[31] Such information asymmetries are likely to get worse, as more information from the internet of things goes live, significantly increasing the scope and quality of personal information that might be repurposed.

And there is little that IT managers can do about it. While they can send advance warnings informing users within the organization about such attacks, they can do little to protect users working from home or their families. They can do even less to protect devices that are shared with others at home, that are used outside work networks, or that are used by a vendor or anyone in the vendor's extended network. Finally, they can do virtually nothing to assess what information users, even those within the organization, are inadvertently leaking, what information about them has already been breached, or what information is likely to lead to a future attack.

While IT departments are limited in what they can do, hackers are only limited by their motivation to look for information and their ingenuity in repurposing it. Thus, today, the vast amounts of available information and the lack of global laws protecting users heavily advantage the social engineer.

### Knowledge Advantage

The technical know-how required to craft a phishing email is minimal. Access to the internet, a free email account, and the ability to compose an email are all that it takes.

Thanks to the internationalization of computer science education, there is also a global pool of knowledgeable coders creating malware-for-hire.

There are even fully functional phishing kits being sold on the Dark Web. These kits host fake websites of well-known companies, such as PayPal, and provide completely branded form emails with built-in tracking tools to assess the attack's success. All a hacker needs to do for deployment is plug in a target user's email address.[32]

Further enabling them is the worldwide dominance of just a few operating systems, device manufacturers, email services, social media platforms, and online music and video services. Because of this, malware creators can specialize. They can expend time and effort toward creating a virtual copy or spoof of a major global service's website or application and offer it to other threat actors. For instance, finding a single weakness in one of Adobe's software programs or a vulnerability in Microsoft's help desk, given how widely these products are used, can lead to exploits all over the world.

The breach into Yahoo's email service best illustrates this. The single breach, effected by a Russian hacker, compromised emails of over a billion users all over the world who not only had their email content compromised but also handed over access to all the online services they had authenticated using their Yahoo credentials.

The knowledge advantage of hackers is further enhanced by their access to potent malware stolen from national security agencies. A case in point is the infamous WannaCry and NotPetya distributed ransomware attacks that crippled millions of computing systems worldwide in 2017,[33] where hackers repurposed EternalBlu, an exploit they had stolen from the US National Security Agency.

The hackers' arsenal is also being enhanced by open-source artificial intelligence programs and algorithms. They make it possible for hackers to mine stolen user information, find patterns, guess passwords, and develop phishing attacks that continually adapt to internet users' behavior. Such adaptive attacks can be highly personalized to each user, making it hard for users and IT departments to detect and thwart them.

Another factor shifting the knowledge advantage in favor of hackers comes from a lack of global cybersecurity standards governing hardware and software development. This has fostered poor coding practices and, coupled by a rush to market for products, has led to the never-ending discovery of weaknesses in existing computing systems. Hackers can use any of these to develop exploits and access points.

In 2019, two major flaws in Intel's microprocessor were discovered, Spectre and Meltdown; a second major vulnerability in Apache Struts; a new weakness in USB-C hardware; and 43 critical weaknesses in Adobe's Acrobat Reader and Flash. This doesn't include the dozens of flaws identified in lesser-known online systems and services, and the millions more that remain unknown. Some of these flaws have patches; some others, especially in many IoT devices built by low-cost manufacturers and companies that don't traditionally develop software, cannot be patched; and still others remain unpatched because most users are oblivious to the need to apply patches. Thus, today there are many available vulnerabilities and loopholes in almost every layer of computing technology—far too many for any single IT manager to guard against.

Finally, most in IT have little interest in understanding, let alone dealing with, users. IT managers' backgrounds in engineering and, because of a decade of outsourcing, their reduced control over users have made them less willing and able to deal with users. In contrast, phishers (like Baratov, who helped the Russians in the DNC hack) are skilled at understanding users. They spend time learning about users by conducting different attacks and examining their relative success. They can do this repeatedly because there has been little done to stop them. Hence, the technical arsenal available to today's hackers is limitless, as is their motivation and ability to keep attacking.

### Information Environment

The information environment also heavily favors hackers. Organizations function like the princely states in the Indian subcontinent that allowed the five-hundred-year reign of the Thuggees we met in chapter 1. Most IT departments focus internally on their organization's security, and the impetus is to contain news about security breaches and attacks rather than share it. While there have been moves toward creating sector-based information and analysis centers (ISACs) and law enforcement–led information-sharing groups, organizational participation in them remains voluntary.[34]

Organizations also are secretive about breaches. This is because of the blowback that follows news of a breach. As we saw in the introduction, many of Sony's top management were fired. Such punitive measures exacted by the media instead of by the courts create a preference for concealing information about breaches rather than being transparent.

Finally, breach notification laws governing who needs to be notified about a breach, when, and how vary by state, region, and nation. Thus, whenever there is a breach, the legal ramifications, coupled with concerns about the media's coverage, lead organizations to be reticent to share information about cyberattacks, even with their users, who remain oblivious and open to further attacks.

But because of such policies, a social engineer can use ransomware on one organization and subsequently reuse the same attack on another organization with neither warning the other about it. This came to light in a spate of ransomware attacks that in April 2019 crippled all of Greenville, North Carolina's city government computers. In May, the same attacker targeted Baltimore, Maryland, crippling over 10,000 computers and affecting its billing systems, municipal departments, hospitals, and airports.[35] In spring 2020, this hacker targeted various parishes in Louisiana,[36] and has continued to do so throughout the nation in 2021. In this way, an attack can keep co-opting organizations.

On the flip side, even among organizations that wish to inform law enforcement about an attack, there are limited avenues for doing so. For one thing, there is no central portal, no single law enforcement agency or cybersecurity command center, that can help with all facets of a breach.

Second, the science of cybersecurity lacks a unified ontology, as in a shared language and definitions for key terms. Because of this, even the most basic elements of reporting are still unclear. For instance, it remains unclear what a data breach entails or how it differs from a hack. There is also no scientific consensus on when a hack or a breach has occurred. Is it when the hacker finds information about users to co-opt? Is it when the hacker actually sends or the user receives a spear phishing email? Is it when the user opens the spear phishing email? Or is it when the malware enters the network?

Third, the science of cybersecurity focuses more on the risk from software and hardware weaknesses than on user risk. Thus, there is little to no guidance for IT managers on how to quantify human cyber risk or on how to develop IT security policies that reduce user risk.

Finally, warnings about impending phishing attacks and protections against them don't cover all users. They are inaccessible to many consumers, such as those working in small businesses, students, and retirees. This means an attacker can find an unaware host within a retail store offering free Wi-Fi or within a member of the household who shares the home network and locate devices, even those that are ring-fenced by organizational IT security.

Without the centralization of information, a shared language, and the ability to enact universal protections, the information environment favors those who wish to do harm rather than good.

### Incentive Framework

Finally, the incentive framework favors hackers. Hackers can collaborate, outsource, share their knowledge in Dark Web forums, learn from each other, and attempt any attack. Even if an attack fails, the cost of failure is negligible. Thanks to digital currency, encrypted message and file-sharing apps, and online payment portals, hackers can not only get paid for their crime but can also even launder and clear the proceeds without a trace. The media, which at times presents those who get caught hacking as activists and enthusiasts, also ends up encouraging more people to become hackers.

In contrast, because the information environment motivates secrecy rather than disclosure, most organizations are on their own, reinventing the proverbial wheel of security policies and solutions. The legal ramifications and the losses from breaches have led to increased scrutiny of IT management by lawyers, cybersecurity insurers, and policy makers. This has made IT managers reluctant to try bold solutions, lest they invite even more scrutiny.

Making things worse are the varying guidelines on what IT managers must do to protect users in the enterprise. Take the case of the widely accepted guideline on password complexity and expiration developed by the NIST in 2004.[37] This guideline, which many organizations, such as Microsoft, widely implemented, required that users come up with long and complex passwords, which they were asked to change every three months. To meet the constant requests for new passwords by multiple online portals, users often reused old passwords so they could remember them.

Hackers also recognized the quarterly password change trend and began sending password-reset phishing emails to users, which many users, routinized by such communication with IT, complied with. It was not until 2017, more than a decade after countless organizations had implemented the guidelines, that the NIST finally reversed this policy.[38] However, to this day, many organizations and many online services continue to ask users to change their passwords every few months.

Finally, an often-misjudged aspect of hackers is their sophistication. Until the spate of state-sponsored hacking by the Chinese and Russians, hackers were often characterized as pimpled geeks working out of a basement. Many of today's hackers are extremely well organized. For instance, in 2018, a group of hackers had created a subscription service through which they were selling prepress releases from PR Newswire's service, stolen using spear phishing, to investment firms and brokerages. The group had made over $100 million over a five-year period.[39] Much like any organization would spend on R&D, the hacker group was reinvesting some of the capital into the development of even more sophisticated spear phishing techniques. In contrast to this, in most organizations, IT security is seen as a cost center, not an investment. Its budgets are hard to justify and thus are frequently cut.

With limited budgets, varying guidelines, and reduced control over organizational IT decisions, IT managers are motivated to implement the broadest, least likely to disrupt, legally permissible, and widely adopted cybersecurity initiatives. Meanwhile, hackers are constantly upgrading their skills and evolving increasingly ingenious attacks, so doing what every other organization is already doing only makes it easier for hackers to apply the same successful exploit across many organizations.

Solving the social engineering problem of cybersecurity requires that IT managers think differently. The need for this is rather urgent. As chapter 3 will discuss, the current method isn't working.

# 3 HOW CISOS ARE DEALING WITH SOCIAL ENGINEERING

There is something in common among nineteenth-century travelers who died at the hands of Thuggees, those who sank on the *Titanic* in 1912, those boarding the *Hindenburg* in 1937, and even those flying on transatlantic jets in the 1960s—and it's not what you think.

It's that they all carried their belongings in unwieldy haversacks and bulky suitcases. They had to—because luggage with wheels hadn't been invented yet. It was not until 1970 that the president of U.S. Luggage put wheels on a new line of suitcases.[1] By then, we had broken the sound barrier, been to the moon, and even sent text messages over the ARPANET.

Why did it take so long for such a seemingly obvious innovation to be realized? After all, wheels had existed for millennia, people had traveled for just as long, and everyone always carried an assortment of things with them. Yet, luggage makers the world over were locked in their ways of thinking about the problem.

This is not just the case with everyday inventions; science, too, suffers similarly. Most of us are familiar with how for over 1,500 years scientists rejected the idea that the sun was the center of our universe until Copernicus and Galileo corrected this misconception in the sixteenth century. There are many other examples of wrong scientific ideas, from diseases being caused by noxious air to the universe being full of luminiferous aether, that prevailed for centuries. Scientists were caught in a pattern of thinking—a paradigm.

In his seminal book *The Structure of Scientific Revolutions*, Thomas Kuhn attributes such paradigms to the process and practice of science.[2] Science, while objective in its individual application, is an activity conducted by a community. This community of scholars, thinkers, educators, and practitioners frames each phenomenon they are working on. They provide the language to organize it, the data to support it, and the arguments against contrary views. This framework is imparted to students, who become its practitioners and its proponents. This shapes not only how phenomena are approached but also how the problems in the field are viewed. Over time, it is hard even for experts in the field to recognize they are trapped in a paradigm, let alone think outside it.

But this isn't some phenomenon that only affected early scientific thought. Contemporary scientists still get locked in paradigms. Take a recent discovery that changed the science of archaeology. Since the sixteenth century, archaeological research on ancient Greeks and Romans viewed the gods, giants, griffins, and cyclopses depicted in their classical mythology as the product of a comparatively primitive people's vivid imagination. One person, folklorist and scholar Adrienne Mayor, changed this view.[3] Her research showed the creature myths were rooted in the bones of prehistoric animals whose fossils abounded wherever such legends emerged.

We now recognize that the ancient Greeks were fossil hunters who were collecting excavated bones and interpreting them using their limited understanding of animal anatomy. So, the skull of an ancient elephant-like species, deinotheriidae, which had a large central cavity for its proboscis or trunk, gave rise to stories of the giant one-eyed cyclops. Extinct giant giraffe bones inspired the Homeric legend of Heracles rescuing Hesione. Fossil skeletons of mastodon, protoceratops, and other prehistoric animals led to ideas of Dionysus's giant war elephants, gold-guarding winged griffins, and bellowing monsters.

But archaeologists have had evidence of this all along. Not only had the bones of prehistoric animals been found in these areas, but the ruins of many Greek temples also contained relics that were discovered during digs. However, because of their worldview, archaeologists treated them as accidental discoveries and discarded them.[4]

Today, computing is similarly locked in a paradigm: the engineering paradigm. All its core principles—from how it defines problems to what solutions it seeks—are driven by the engineering disciplines.

A telling example is in the terminology used by computer engineers to signify errors in software that result in unexpected or inaccurate results: a bug. Bugs in software are almost always caused by human errors during software programming. Rather than admit this—and, as some have argued, call it "a blunder"[5]—the field frames the problems as organic aberrations that like pesky little critters just show up in software. This influences not just how the field frames the problems but also solutions like debugging, which focuses on the technical process of finding and rectifying such glitches by using automated processes and debugging tools rather than focusing on improving human coding.

The dominance of the engineering view in the computing space stems from three levels of influence. First, at its core, computing devices are made of electronic circuits, a traditional area of study for engineers. Second, the proponents of the field, whether they create hardware or software, have a proclivity toward electronics and technology. They tend to be trained in engineering programs that focus on physical science principles, where the outcomes are predictable and deterministic. They focus less on human processes, which are unpredictable and stochastic. These views shape their understanding of technology and their expectations from the world at large. Third, the field is organized into disciplines such as computer science and computer engineering through which they socialize students and practitioners in the ways of the field and the ways of approaching problems.

Because of this, the field approaches the problem of social engineering primarily as a technical issue, one that can be fixed with software or hardware. IT managers who are trained in engineering demand deterministic solutions. There isn't necessarily something wrong with this. Nor is there anything wrong with engineering as a field or engineers as a class. Engineering has made enormous leaps in human progress, and engineers have helped actualize these through designs and devices. But it is one thing to expect machines to behave predictably and another to expect people to behave like machines, and therein lies the problem. The engineering orientation toward computing technology has shaped the view of the problem and the expectations from its solutions. It has led to a bias toward technological solutions that are predictable and understandable rather than human-focused interventions that are unpredictable and unfamiliar.

A vast solution space has emerged to fill this need. It is dominated by software and hardware products, each claiming to be the "silver bullet" that can resolve the problem of social engineering. Most of these products either completely ignore users, providing internal data and reasons for it, or incorporate them peripherally, expecting users to provide only a few trained clicks on some software's interface. All cybersecurity engineers portend the day—which is always said to be around the corner—when even this won't be necessary, when technology will completely outengineer the user from falling for social engineering or make users entirely obsolete. Consequently, there is limited interest or value seen in understanding users.

Such views have inspired two broad approaches to combat the user issues that social engineering exploits. One takes an overtly technical approach, focusing on hardware and software to solve the user problem. The other has origins in the industrial age of shop floors and conveyer belts—the human factors approach. It has led to the security awareness and user training initiatives that are implemented by organizations the world over. This chapter presents the inherent strengths and weaknesses of both approaches, with a particular emphasis on the effectiveness of user training because of its wide acceptance as a proactive approach to creating user resilience.

### THE TECHNICAL APPROACH

We can broadly classify the engineering approach to solving the threat of social engineering into two Cs: containing and constraining. Each C refers to a method or technique for achieving cybersecurity. Each aims to control access to technology, just in different ways. Containment techniques focus on keeping the hackers or the bad guys out and keeping the good guys, meaning the authorized users, inside the network. Constraining methods focus on creating siloed systems, compartmentalizing them in various ways so as to limit unauthorized access to them. Often, there is overlap in how individual solutions achieve their objectives. Thus, the classification into different C's is for interpretive convenience.

Containment and constraining solutions almost always involve software or hardware. They emerge from research and development in computer science and engineering, where there is a bias toward such solutions. They are consequently the most common classes of solutions sold by cybersecurity vendors. Rather than produce an endless list of all of them, this section presents a top-level view of some leading solutions that espouse these techniques and how they contain or constrain.

The most popular techniques for containing access include air gapping, firewalls, and sandboxing. Air gapping involves isolating computers from networks. The goal is to make it hard for hackers to gain access to other computers on a connected network by compromising any one device on it. Air gaps can be created physically by isolating a computer or series of computers from others. For instance, email use can be air gapped by restricting it to computers that don't have access to any files or servers within the organization. It can also be achieved using software, as is more commonly done. Here, a program electronically separates some computers from others. As a technique, air gapping is commonly used to protect high-value information from the general computers that access the internet. They are also used to isolate and protect control systems such as those managing industrial systems, nuclear facilities, electric grids, hospital devices, and airplane flight controllers.

Another technique for containing access is through firewalls. This is an electronic method of network isolation and involves controlling data traffic coming in and out of a computer. Firewalls place a barrier between networks and computers and can be positioned on the gateway into the network, monitoring overall traffic, or on host machines, protecting just individual devices. They can also be configured to route traffic, block questionable pages or packets, or monitor specific applications and file types.

A third commonly applied method of containment is sandboxing. This creates a virtual copy or proxy of a computer or even an entire network so that applications run in that virtual version rather than the actual system. By executing software in a restricted virtual environment, sandboxing prevents any infection from affecting the overall computing system. This process can also protect core system processes from being accessed by other programs, thereby maintaining the integrity of the overall system. This is how Apple's "walled garden" works. It is a virtual sandbox that protects its core operating system from rogue programs—a reason for the iPhone's high reliability.

Similar to containment techniques, there are a host of popular constraining approaches, starting with various identity management systems, as we discussed in chapter 2. These include the traditional login- and password-based authentication system and the various multifactor authentication systems (2FA, physical token, biometrics, and such) layered over it that limit access only to authorized users. Other mechanisms for constraining not covered previously include the use of virtual private networks (VPNs), whitelists, and virus protection software. VPNs encrypt communications between computers over a network to restrict the ability of unauthorized persons to capture the data being exchanged during transmission. Whitelists are preapproved lists of programs, websites, or email addresses that are permitted access to a system. Anything not on the list is automatically blocked, constraining user access.

Finally, constraining can be achieved using virus protection programs that restrict or remove viruses, pop-ups, malware, or trojans. Older forms of such software captured the signature of viruses and developed remedial patches that were sent as updates to the virus software's firmware. Newer virus protections incorporate machine learning to find common patterns in malware. Some use behavior-based identification approaches, where the actions of programs are tracked to predict whether they are deviant, while some others use a heuristic approach, where the program's software code is examined to predict its malicious intent. Besides these, there are many other techniques, ranging from the highly technical to the simple, that contain or constrain.

At the more technical end are various network monitoring techniques that track all data flows within a network and build a user-behavior profile, which can help detect anomalous activity, and email scanning software, which can look for questionable keywords and technical clues to the deceptive intent of emails. At the lower, nontechnical end are approaches that block certain types of connections on devices, disallow certain websites (such as those lacking specific types of encryption), curtail the display of hyperlinks or images within email, and block emails that are not from approved sources from being delivered.

## ADVANTAGES AND LIMITS OF THE ENGINEERING APPROACH

The entire swath of solutions aimed at containing and constraining access have several advantages, chief among them being that they involve a technology that predictably and demonstrably works. Be it air gapping, firewalls, sandboxing, biometrics, multifactor authentication, or antivirus software, it is easy to see how they work on a computing network through "red-team" attack simulations. Thus, it's also easy to make a case for implementing them.

Another advantage of both approaches is the control they provide IT departments over the systems under their purview. Containing solutions limit the possibility that unexpected events will cripple the entire enterprise's IT systems. Constraining solutions likewise help remove those users who don't need access to critical system processes and resources. In doing so, they reduce the variability in adverse IT-related events that could occur.

Each solution also enhances the effort and investment required from hackers. By placing various technical hurdles, they make it harder for attacks to succeed. This makes the organization a less desirable target. Besides, if a system gets compromised, technical solutions, because of the access and operational silos they create, help quickly identify the location of the failure, which helps in providing speedy remediation.

External security vendors can also manage many of the solutions. This appeals to organizations, which increasingly have limited in-house IT staff to support them. These solutions can be implemented with minimal user involvement, often without ever requiring that the user do anything on their computer. This makes it easier to deploy the solutions to users and keep the computers patched and up-to-date. Solutions such as virus protection, firewalls, and pop-up blockers are also accepted as necessary facets of users' computing experience on many operating systems. They face little user resistance during implementation.

Finally, a visible product accompanies each technical solution. This is important for CISOs, who must show that they did something tangible to ensure enterprise security.

Almost all the products that contain or constrain access help do this. Because many of these solutions have already been implemented by other organizations and their CISOs, there is the added advantage of showing not only that something is being done but also that it is what others in the field are doing. With cybersecurity seen as a cost center, having a tangible product that other organizations in the same space are already using makes it far easier to justify the solution.

While these approaches offer many obvious advantages, they also suffer from a host of limitations. For one thing, despite their many promises, no single solution can protect against all forms of cyberattacks. Air gapping can silo networks by preventing rogue programs from being executed, but it also keeps them vulnerable by not allowing protective patches and critical updates to be applied. Similarly, whitelisting might block unapproved connections but cannot discern approved connections that have been compromised. And while virus protections and firewalls can prevent malware infections, they cannot protect against pretexting and other forms of social engineering attacks.

Furthermore, because of the technical constraints imposed by various platforms, many solutions are platform specific. For instance, antivirus, email scanning, and most email marking software is not supported by mobile email applications such as Apple's iOS Mail. Even the solutions that do work across platforms are actually more limited than they appear. Take the case of homographic attacks, where the content of an email or URL has characters from other languages that appear to be English language characters. An example would be replacing characters in the English word *password* with the Cyrillic "o," a minor alteration that is indiscernible to most people looking at it. The problem is that most antiphishing software trained to flag the word *password* in emails also cannot catch this change. It is precisely what the hackers did to evade the Democratic National Convention's phishing email scanning software—and we know what happened because of that.[6]

Second, many of these solutions are limited by the sheer volume of data and the variety of ways users within organizations today access it. IT users include different employees, vendors, and regulatory agencies. Each user interacts with the company's networks at different levels, using different devices, and from different locations. Cumulatively, users in even a medium-sized organization with, say, 50 employees can exchange petabytes of data via email, messaging, videoconferencing, and cloud-based services.[7]

Tracking every one of them is a hard task, made even harder in organizations where privacy regulations prohibit certain forms of monitoring. These include employees in federal government agencies, law firms, health care organizations, and organizations that fall under the European Union's General Data Protection Regulation. Thus, many user interactions within organizations simply cannot be comprehensively screened, tracked, and monitored.

Third, each engineered solution also brings a new set of technologies into the workspace, all of which can be co-opted by social engineers. For instance, regular firmware updates from antivirus software can routinize their downloads by users. Social engineers can mimic or spoof such updates. Users who habitually click and accept such updates then get easily tricked. This was how the North Korean hackers infected the Sony Pictures computing system with their ransomware. Likewise, email marking software that flags external emails routinizes users into looking for such flags, and these, too, can easily be spoofed.

Additionally, many online protections create a sense of invulnerability and make users more cavalier with their online actions. Research points to users of antivirus software being more careless while opening questionable sites and emails on their devices.[8] The same problem plagues users of email scanning software, who likewise come to depend on the system's judgment. This was ostensibly why the spear phishing email sent by the Russians to John Podesta was trusted by the DNC's IT help desk staff, whom Podesta had asked for guidance. The IT staffer trusted the email scanning software in use, which, having been tricked by the homographic replacement of the English character in the word *password* with the Cyrillic "o" in the email, did not flag the email as spear phishing.

On the flip side, some technical protections return false positives, such as flagging authentic email and websites as being fraudulent. This ends up having the reverse effect. They reduce users' trust in the protections offered by these technologies. Over time, users ignore the system prompts, often unbeknownst to IT managers, which opens the organization to compromise.[9]

Finally, all the technical solutions are designed by engineers using limited user-experience tests. They become cumbersome to use in real work environments, where the users tend to utilize the technology in a variety of ways, many of which are unaccounted for during software development. This leads to users finding workarounds that negate their safeguards and expose the system to even greater compromise. Take this anecdote from a security agency in Singapore that was relayed to me. The agency had instituted a physical air gap separating computers on which users checked emails from their main computing system. The idea was to stop spear phishing emails from being ported into their central computers. They got breached, however, when an authorized user transferred an emailed attachment from the email computer to the main network using a USB drive. This ported malware concealed in the attachment with it, and because the main network was thought to be protected, the hacker went undetected.

Take another case from a federal government agency in the US. For fear that users opening websites could launch malware, their IT department disabled all hyperlinks displayed in emails. They, too, later discovered something about email use that should have been obvious: that when users couldn't link to a site by clicking on the hyperlink, they were copying and pasting the hyperlink from the email into their browser, negating the effectiveness of this safeguard.

In each case, users did what all users do with email—use it to do more than just communicate. From document exchanges to cloud-based file-sharing links, a lot of what's accomplished via email requires downloading attachments and launching URLs that are exchanged by email. When deploying such seemingly secure solutions, IT managers never considered this obvious use.

But even many complex technical solutions implemented by IT departments do likewise. For instance, strict whitelists that block certain messaging, social media, and shopping sites disregard the purpose these apps serve. In many organizations, employees prefer to text each other rather than use phone calls. In some nations, websites such as Facebook and Amazon serve as the essential link between people, used even for currency transfers and business-to-business communication, and all this was before the COVID-19 pandemic made sharing URLs to various videocalling portals and cloud services, and using a variety of devices to connect to work servers, necessary for organizational continuity.

In today's dynamic work environments, staying competitive, current, and knowledgeable necessitates that employees have access to social media, news services, videoconferencing, and messaging apps. This cannot happen in the siloed information islands that are premised on workflows that someone in the IT department deems acceptable. Consequently, many users ignore IT safeguards or resort to risky workarounds to access blocked sites—which place the organization at an even greater risk.

Thus, no matter their sophistication, the technical approaches to containing and constraining have their limitations. The biggest limitation is imposed by users and how they interface with, use, and experience engineered security protections. Just like a home security system, all cybersecurity systems require that users be involved in using the system optimally. This requires user awareness and education, which the next engineering approach attempts to achieve.

### THE HUMAN FACTORS APPROACH

During the early twentieth-century era of mass production, people worked on factory shop floors assembling products. The focus of industry was on increasing factory output. The key was to improve operator efficiency and keep the conveyor belt moving faster. To achieve this, the field of engineering began incorporating scientific management practices into production planning. With it came a new view of users: as human factors.

In this view, every worker was seen as another cog whose performance affected the overall speed of the belt. The focus was on what each human did, how this factored into overall production, and how it could be improved. Actions were cataloged using a novel approach for the era, time-and-motion studies, where every operator's actions were tracked against an optimal standard. Those who didn't meet the standard were trained. This was done repeatedly until the desired efficiency in output was achieved.

Nowhere was there a focus on the individual or their motivations. As far as the field was concerned, people's fundamental motive for working was money, so incentives were built offering workers more money for improving production, regardless of the consequences on their life or limb. Their thoughts about work, their capabilities, and even their personal safety didn't matter, because they were getting paid. Some of this changed over the years, as legislation defined limits on hours worked and imposed safety standards. Globalization and increased competition also impacted it. Shop floors were replaced by start-ups and technology-run enterprises, and foremen and daily-wage workers were replaced by entrepreneurs and salaried women and men. However, the term *human factors* stuck, as did some of the foundational views about users.

Today's engineers use the term *human factors* as a catchphrase—in academic papers, policy doctrines, and job announcements—to describe almost all user-related issues in computing. It encompasses the entire body of research on user vulnerabilities and the proactive approaches to combating social engineering.[10]

But, because of its roots, the focus of human factors research in computing continues to be on user actions. Using different approaches, various user behaviors are coded, tracked, and monitored. Some of this data informs technology design, some identifies user issues, and some is collated, forming the basis for all of today's user awareness and training solutions. All these approaches assume human actions are a proxy for all human thought—a fact that was flawed from the earliest days of the industrial shop floor. Not only do most of our thoughts not translate into action, but many of our actions—our habits, routines, and patterns—aren't even premised on forethought. Most of us do fewer things than what we think about and often do many things without thinking.

So, while the computer sciences have plenty of names for software bugs—syntactic bugs, control flow bugs, Mandelbug, Heisenbug,

Schroedinbug—the study of human factors seldom considers human cognition, affectivity, individuality, personality, or habits, all of which have been shown to influence how users encounter, interpret, and use technology. In other words, there just isn't a lot of human science in the study of human factors.[11]

Following its mechanistic view, the human factors approach sees users as inefficient computer operators—squeaky cogs in the network machine—performing a finite number of suboptimal actions that can be identified, enumerated, sequenced, and optimized through hands-on training. Training is the primary, and often the only, solution to almost all user issues.

Within the field of security, the training solutions that have emerged take one of two broad forms. One is more hands-on and involves didactic instruction that emulates classroom-type education. Such didactic security training can cover a wide range of topics, from the basics of email use to the correct ways to apply firmware updates, use VPN software, or navigate the company's approved file-sharing services. At times, awareness campaigns advertising how to be safe online are used to supplement the limited reach of the instruction. Most such training culminates in quizzes and tests, with some even incorporating games and other entertainment approaches to engage users. Many provide completion certificates that employees are required to renew periodically.

The other training approach that has evolved involves the use of simulated phishing attacks called pen tests. The approach assumes that for training to be more salient, users need to be shown how vulnerable they are. Users, usually those whom the IT department sees as posing some risk to the enterprise, are therefore subjected to the pen test. In most cases, the pen-test email has a hyperlink or attachment in it. Anyone clicking on the link or attachment is seen as having failed the test. The system then pops open training that is embedded in the hyperlink or attachment, which shows the user what they missed in the email that could have revealed the deception. By design, such embedded training requires frequent pen tests, each employing different emails, sent at varying intervals, to cover a wide swath of social engineering attacks and also to keep users always alert to the possibility of an actual attack.

Both didactic and embedded cybersecurity training approaches are increasingly being implemented in organizations all over the world. Part of the reason is that there is no solution other than training to educate or inform users about cybersafety. The other reason is that awareness training is increasingly being mandated. In the US, the Federal Information Security Management Act requires that all users in federal government organizations and in any organization working with them (including contractors, foreign or domestic researchers, visitors, guests, or other collaborators) complete cybersecurity training. Many individual states in the US and nations across the world are also passing similar mandates,[12] because of which training today is the leading human factors–focused approach to improving user resilience against social engineering.

## ADVANTAGES AND LIMITS OF THE HUMAN FACTORS APPROACH

There are many good reasons for training users, beginning with the activity of training, which appeals to the basic worldview of engineers. After all, it is the method used to instruct different machine operators, from automobile drivers all the way to astronauts. So why wouldn't it help train users on cybersafety? Following this logic, embedded awareness training is considered more effective, because it is more hands-on and, much like shop-floor training, can be applied continuously at the user's desk, unlike didactic classroom-type instruction, which can only be done at discrete intervals.

Training also comes highly recommended. For instance, the National Institute of Standards and Technology's Cybersecurity Framework, a widely accepted blueprint for how organizations can assess and improve their cybersecurity posture (discussed further in the next section), endorses cybersecurity awareness training for building user resilience. Much of today's training comes in packaged form, as a technical product. This brings all the many advantages that we already touched on when discussing the engineering approach to security.

There is a further advantage: user data. Data from training packages includes users' training completion rates, their time spent learning, and—among the most important indicators—data on who failed the simulated phishing pen tests. The failure data serves as a quantitative metric of the overall cyber risk posture of the organization. For instance, the US Department of Homeland Security recently instituted the Continuous Diagnostics and Mitigation (CDM) program—a dashboard that provides tools for assessing cyber risk in federal government organizations. A key CDM output is a cyber hygiene score for each organization. It is partly derived using the pass-fail data on embedded security training. Thus, training has a number of advantages, from being recommended to being a product that can provide reams of data, which can serve as inputs for assessing the organization's overall resilience.

But for all these advantages, there are also several disadvantages. For one thing, the activity of training, especially if mandated, can quickly turn into an act of compliance that every IT manager engages in merely to tick it off their list of required security initiatives.

Second, when training data becomes part of the organization's cyber hygiene score (as the CDM discussed here does), it can devolve into an anxiety-driven exercise both for IT managers and for users. IT managers might keep increasing the frequency of training to reduce their overall risk scores. This can lead to punitive actions, which some IT managers have already been doling out to users, in order to scare users into compliance. On the flip side, IT staff, responsible for lowering the score, might craft tests that are easy to detect to improve the overall detection rate.

There are other contaminants that frequent, forced training might inject. In surveys I have conducted in organizations, employees have complained about "training fatigue"; supervisors, to improve their team scores, have forewarned their employees of a phishing test; and many employees have avoided opening company emails on days when they know a test is imminent. Such factors not only go unmeasured but also detrimentally affect employee morale. More importantly, they undermine the validity of phishing test results and the resultant cyber hygiene scores that the organization relies on.

Third, there is an upper limit—a ceiling—to how much knowledge user training on cybersecurity can impart. Computing technology is highly complex and multifaceted. Vulnerabilities are innumerable. No single person, not even the brightest computer scientist, can claim knowledge of all technical and nontechnical exploits on any of today's computers. Thus, training employees to become completely cybersafe is an impossible goal. Given this, most training programs make a trade-off between what they believe are pressing vulnerabilities and the users' likelihood of encountering them in their job. These decisions are based on a priori assumptions about known cyberattacks and the skills that engineers believe people should possess. This then becomes the topic or content area of instruction. Such approaches not only completely ignore users' task-based training requirements but also often lead to a program that has no appeal to the user.

Furthermore, the variance in programming of individual operating platforms and functions makes all training limited in scope. Take the simple case of spear phishing awareness training, where users are taught to check an email's message header. Email headers contain snippets of code that reveal any incoming email's domain name, the sender's IP address, and the email's true source. Users can assess whether the email is authentic or a spear phishing spoof by reviewing it. While it is good to ask users to do this, it's hard to perform this in practice even on a single device, let alone on many different devices. Take the iPhone as a case in point. Its default Mail program doesn't display email headers, while the Gmail app available for the device does. So any user trained in this cybersafety skill would either have to switch programs or switch devices each time—for each of the multitude of emails they receive each day.

Such platform- and software-specific differences permeate users' computing experience. Different browsers have distinct ways of presenting secure socket layer (SSL) information, which reveals the presence and quality of the website's security certificate; browsers handle homographic domain names in URLs (the display of domain names with characters from other languages, such as using the Cyrillic "o" to create a fake Google web page) in different ways; and email programs have various ways of handling graphical images and tracking cookies. These variances stem from how individual technology companies approach security, and they persist and evolve with platforms. They also change as software is updated, when new functions are added, and as new national or international regulations force changes. Because of this, all training has a very limited shelf life—one that is dictated by various applications, operating systems, computing devices, and by how, when, and where they are used.

Finally, most training programs rarely consider how users think about cybersafety. For anyone not directly working in IT security, cybersafety is an afterthought—something they seldom think about. This is because in organizations the world over, users are rewarded for being efficient and effective, and seldom for being cybersafe. Yet, training approaches expect constant vigilance, even proactive behaviors, such as always reporting suspicious online events to IT managers.

Most training programs also advocate cybersafety routines that just ignore how users behave in the real world. As a case in point, one spear phishing awareness program trains users to look at dozens of different facets of an email in order to ascertain its veracity. Such actions are hard enough for any user to engage in while examining a single email, let alone when they—as most of us today do—receive hundreds of emails in multiple in-boxes throughout the day, which they often respond to while on the go, using different devices that vary in screen size, operating environment, and visual layout. Expecting users to focus on so many details for each email is unrealistic and reflects an operator view—a paradigm rooted in the engineering tradition of viewing users as automatons who engage in repeated, enumerable tasks and who can, with just enough repetition and instruction, be trained to perform better.

## THE EFFECTIVENESS OF SECURITY AWARENESS TRAINING

Thanks to its many advantages, security awareness training is among today's most popular proactive interventions applied by organizations to prepare users against social engineering. To demonstrate how user training is planned for and achieved, figure 3.1 shows Intel Corporation's plan to address the cyber risks posed by its employees. Intel applies the highly regarded NIST Cybersecurity Framework that I mentioned earlier in the chapter. The training plan shown in the figure is from the NIST website.[13]

For those unfamiliar with the NIST Cybersecurity Framework, here's how it works. The framework is designed to help organizations define their

| Customized tier definitions | | | | |
|---|---|---|---|---|
| Focus area | Tier 1: Partial | Tier 2: Risk informed | Tier 3: Repeatable | Tier 4: Adaptive |
| People | • Cybersecurity professionals (staff) and the general employee population have had little to no cybersecurity-related training.<br>• The staff has limited or nonexistent training pipeline.<br>• Security awareness is limited.<br>• Employees have little or no awareness of company security resources and escalation paths. | • The staff and employees have received cybersecurity-related training.<br>• The staff has a training pipeline.<br>• There is an awareness of cybersecurity risk at the organizational level.<br>• Employees have a general awareness of security and company security resources and escalation paths. | • The staff possesses the knowledge and skills to perform their appointed roles and responsibilities.<br>• Employees should receive cybersecurity-related training and briefings.<br>• The staff has a robust training pipeline, including internal and external security conferences or training opportunities.<br>• Organization and business units have a security champion or dedicated security staff. | • The staff's knowledge and skills are regularly reviewed for currency, applicability, and new skills, and knowledge needs are identified and addressed.<br>• Employees receive regular cybersecurity-related training and briefings on relevant and emerging security topics.<br>• The staff has a robust training pipeline and routinely attends internal and external security conferences or training opportunities. |

**Figure 3.1**

Intel's plan to address user cyber risk.

*Source:* NIST.

overall cyber risk posture and plan for achieving cyber resilience. It has three components: a core, where the organization lays out a top-level strategic view of its goals and activities; risk profiles, where the organization outlines the current state of its cybersecurity activities versus its desired state; and implementation tiers, where the organization lays out the specific activities it intends to engage in to achieve the objectives set in the core within each functional area of concern.[14]

The framework further specifies four implementation tiers. In each successive tier, the organization is expected to achieve a greater reduction in cyber risk through security interventions that increase in rigor and sophistication. Tier 1 is when an organization is partially informed about risk, usually because it has ad hoc cybersecurity policies. Tier 2 is when the organization is more informed about its cyber risk because it engages in a defined risk-assessment process. Tier 3 is when it has a more formal risk management practice that is regularly communicated and updated. Finally, the highest tier, Tier 4, is when the organization's risk management processes are agile and adapt to changing threats. Of the four, Tier 2 is what is minimally desired, while Tier 4, because it is expected to achieve cyber resilience, is optimal.

In the figure, Intel details how it plans to graduate to the NIST framework's highest risk management tier in preparing its employees against social engineering. In Tier 1, Intel identifies the lack of cybersecurity awareness among its employees and the lack of a training pipeline. In Tier 2, Intel plans to make its employees risk informed by using general awareness training. Next, Intel plans to enter Tier 3 through the implementation of regular training and briefings and by establishing a "robust training pipeline." Finally, through regular cybersecurity-related training and briefings and by again establishing a "robust training pipeline," Intel plans to become risk adaptive, thereby achieving cyber resilience in its people focus area.

While Intel doesn't specify the "robust training pipeline" it plans to implement, one thing is certain: it plans to use training—and a lot of it at each tier—to improve employee awareness. In fact, the words *training* or *awareness* appear a dozen times in its plan. Given that only two approaches to training (didactic and embedded) exist, Intel's training pipeline more than likely involves some didactic training and many repeated embedded phishing tests. Surely, one of the world's largest semiconductor manufacturers, with a market capitalization in excess of $200 billion and experience with over 100,000 employees, can do more than constantly train its employees. Think about it this way: Intel spends close to $10 billion each year on research and development, which is performed by people—the employees working at Intel. All the company's intellectual property, product blueprints, and future development cycles could be compromised using social engineering. A single spear phishing email could accomplish this. And Intel's only antidote to it is user training.

But there is another facet of their plan that is noteworthy. Intel's Tier 1 begins by defining their user problem as the lack of a training pipeline in cybersecurity. Their plan begins by assuming the absence of the solution as the problem. Without so much as diagnosing users, let alone considering whether or what cybersecurity training approach is necessary or effective, Intel assumes this lack of training among users is the reason for their cyber risk. And it's not just Intel. I am not picking on them. They are merely a convenient example, available on the NIST website, that helps illustrate the prevailing axiom: *Users are vulnerable because they lack cybersecurity training, and training reduces this lack.*

Following this, organizations from Wall Street to Main Street have been blindly applying training to reduce their users' cyber risk, with organizations planning to spend close to 40 percent of their cybersecurity budgets on security awareness training in 2020.[15] According to a Rand study, by 2023, such costs are expected to increase exponentially—by 400 percent—outstripping even the losses from cyberattacks, which are expected to grow by about 200 percent.[16] This exploding demand has spurred a cottage industry of training vendors, from small, home-based businesses to multimillion-dollar corporations, with even traditional technology companies such as Cisco and Microsoft looking to enter the space. Already, the vendor space has witnessed tremendous growth in investor interest, with training companies receiving billion-dollar valuations and many others receiving millions in seed funding.[17]

But even as training is being blindly prescribed and its providers are benefiting, there is a pivotal question that remains unanswered: how effective is all this training in preparing users against social engineering attacks?

The answer to this question is rather difficult to find. This is because objective data on training effects across various organizations and sectors is not publicly available. The reason for this is the nature of competition in the cybersecurity space. Most companies are very secretive about their research data, with less than 15 percent of cybersecurity research studies that gathered original data being made available to other researchers.[18] This stems from intellectual property considerations, privacy concerns, and fears of negative publicity and regulatory oversight (an issue that companies such as Facebook learned the hard way when they made their internal experiments on emotions known to the public).

My experience dealing directly with security training companies has been similar, with most being unwilling to subject their products to independent assessment, let alone allow access to their own research data. Organizations that use training are reluctant to share their internal data for similar reasons—fear of negative publicity, fear of attracting regulatory oversight, or, worse yet, the fear of being seen as vulnerable by hackers. Thus, there is little objective data available from cybersecurity training companies on the effectiveness of their training.[19]

So, we have to rely on alternative sources. A solid stream of indicators comes from research studies conducted by academic and research institutions. Many of these utilize controlled experiments on different users, giving us an understanding of how well training works on different user populations. Among the earliest was research conducted in 2004 by the US Army Cyber Institute at West Point, New York. In these experiments, collectively called the Carronade studies, five hundred army cadets were randomly selected from various classes, given phishing awareness training, and subsequently sent a phishing pen test to assess readiness. Almost all cadets—80 percent (or four hundred)—even after four hours of computer security instruction, fell victim to the phishing test.[20]

In another study, 10,000 New York State employees were phished in 2005. Of these, 17 percent clicked on the link and 15 percent began entering personal information. Despite receiving reprimands from their IT managers, when phished two months later, 15 percent of the employees fell for another attack.[21] In yet another study, in 2011, the federally funded research and development center MITRE tested the effectiveness of embedded training on 1,500 employees randomly selected from a midsized organization based in Washington, D.C., by phishing and then training those who fell for the attack. Across a series of trials, the research found anywhere from 35 percent to 80 percent of employees clicked on simulated phishing email hyperlinks regardless of whether they had been trained.[22]

These are among just a handful of research studies that exist. A recent meta-analysis (a method that aggregates research findings from across different studies) found just 12 published research articles examining training over the course of 16 years. Of these, just four compared the effects of training over time—and their findings couldn't be more equivocal.[23] One showed that the training effect disappeared in four weeks, another that it disappeared in two. Yet another found no immediate or short-term effect, while the fourth study showed training wasn't effective after 10 days but was suddenly effective after 63.

There is yet another line of evidence that points to the relative ineffectiveness of user training. This comes from continuing media reports of cyber breaches and hacks because of social engineering. Within a few months in 2020, ransomware attacks crippled city government servers and systems in North Carolina, Maryland, Georgia, Texas, and Louisiana. All used spear phishing to activate the compromise. Many were in organizations that received mandatory cybersecurity awareness training, not to mention those where the various approaches to containing and constraining discussed earlier in this chapter were already in place. Such reports aren't limited to companies in the US. There are similar reports from all across the globe.

The data from all these sources questions the prevailing axiom that training improves phishing preparedness. But there is data from one additional source. This comes from my experience conducting phishing tests and from the emails I have received from IT managers and CISOs of banks, federal government agencies, and organizations from various parts of the US, Canada, France, the Netherlands, Singapore, India, Australia, New Zealand, and Zimbabwe. Almost all reached out to me after they implemented various forms of training—some over a few years—that failed. So, if you are reading this and have experienced similar results from your training approaches, you are in good company.

Today, most organizations, regardless of how much capital or how many people they have, prescribe the medicine they have at hand—training—because they axiomatically believe that the problem is users' lack of security awareness because of a lack of user training. It is analogous to a patient walking into a doctor's office and being told they suffered from not taking the medicine that was on the doctor's table. Without so much as considering whether training is necessary, let alone assessing who needs it and how much of it is needed, everyone today is then prescribed the medicine: training. Worse yet, organizations dole out the same training to everyone repeatedly, and if the users don't improve, organizations blame them and call them the problem, then dole out even more training. Thus, the "people problem" of cybersecurity is more than a description of a condition; it is prophetic.

The starting point for any medical treatment is an understanding of what ails the patient, but this pivotal step has been skirted by security engineers who think of users as machine operators. They have never stopped to consider why people fall victim to social engineering—the crucial factors that can help us accurately diagnose and treat them. Chapter 4 begins where we should have all along, with an explanation of why users fall for phishing attacks.

# 4 WHY DO PEOPLE FALL FOR SOCIAL ENGINEERING?

"Gotcha!" The word flashed on the screen each time someone clicked on a simulated phishing attack or pen test at this organization in the US. It was a unique touch from National Accounting Agency (NAA) chief information security officer (CISO) Brian Marks, meant to shock the user into remembering that they had been had.

Marks, a retired naval officer, brought the military style of training, where cadets were trained repeatedly with training that was both memorable and consequential, to the workplace. He particularly favored the embedded (using phishing pen tests and then training) approach because it was both. It highlighted phishing failure, making it hard for the user to forget that they had fallen for the pen test, and it was consequential. In Marks's organization, failing a phishing test resulted in a personal email from him and from a supervisor. Fail twice and the monthly newsletter mentioned the employee by name. There was more training online assigned to every offender and some extra didactic training for repeat offenders. At that point, Marks felt most employees would fear seeing the word "Gotcha!" ever again.

Marks's IT team liked his approach. They enjoyed crafting simulated emails that could get users to fail. It became a sort of game that they played. Each month, the team scoured the internet looking for different attacks and then re-created them. At times, the attacks were so good that even Marks's team couldn't help but wonder how anyone could detect them. But many did. In the two-year period since the "Gotcha training" was implemented, overall phishing clickthroughs, as in failures, were down to less than

1 percent. This compared to a 20 percent failure rate in the first month. Clearly it was working. And the fact that even the more difficult pen tests were being detected was evidence enough to the IT team that users were getting increasingly better.

But there was one thing that vexed Marks. There were always some users—a different few—who failed the test. That 1 percent was 1 percent too many. He wanted zero—a number that had eluded him in 24 tries. He had heard about my approach and contacted me. My charge was simple: bring the number to zero or tell us why it wasn't possible. I agreed, with the one caveat that I, not his team, craft the simulated phishing email.

I crafted a pen-test email reproducing communication that users routinely send and receive from their internal cloud-based file-sharing portal. The email wasn't an exact reproduction of the system. Instead, it incorporated just the layout and colors and even had spelling and grammatical errors. The file being shared via the cloud hyperlink, which users were asked to open, was a large PDF document with the Abode PDF icon prominently visible. The pen test was sent on three nonconsecutive days, and each email was tracked to see who opened it, when, how (meaning what device, browser, or IP address), how often, and what they did upon opening it. Two weeks after the attack, all users were asked to answer a few survey questions.

The result of the pen test was that 21 percent of the respondents overall fell for the phishing attack. This was a tremendous difference from the less than 1 percent Marks was used to seeing. In fact, a small percentage of users had even clicked on the phishing email repeatedly.

I had expected this. It's what I have experienced in other organizations where I've implemented this approach. To help explain the results, I had included a few additional questions in the survey. One was a phishing knowledge quiz, where respondents had to identify which of a series of email exemplars were phishing emails. Most of the NAA's users scored near the median (getting about half the answers correct). More importantly, their knowledge levels did not make them statistically more or less likely not to fall for the phishing pen test—another expected result.

Additional questions inquired whether users could spot a phishing test that was being sent in the organization and whether they could anticipate when one was forthcoming. The overwhelming majority stated they could. The repeated embedded training hadn't trained the NAA's users to get better at phishing detection. It had made them better at detecting a test.

There was worse news. Interviews with a random subset of select users who were assured confidentiality revealed that their supervisors were informing them when a test was about to be sent, some users were being helped by others on their team, and almost everyone in the organization had changed their email usage behaviors. This included postponing and rescheduling emails to appear at a different time of the day, which explained the multiple clicks on the email by some users; asking others to confirm whether the email was not a test; opening the suspect email on other devices (smartphones and tablets); or simply ignoring emails, waiting for them to be resent (because their logic was that hackers never did that). These actions were not just reducing efficiency but putting the NAA at greater risk from phishing. Users were just as susceptible, if not more, than before the embedded training program began 24 months earlier, except the NAA's IT team was falsely confident in their users' phishing resilience. So who got who?

The NAA is an actual organization, but I have altered many of the details in the case to maintain confidentiality. When presented with the findings, Marks—much like all the other CISOs confronted with similar results—realized his approach to user training had been completely wrong. He admitted that he never really understood users, but he was ready to learn. He wanted to know why users fall for social engineering. I began, as I always do, by talking about optical illusions.

### SEEING WHAT WE CANNOT SEE

Is this a white and gold dress? Or is it black and blue?

Is that a bird? Perhaps a bunny?

Is that the shadow of a man's face or just a mountain?

Thanks to social media, millions of us have now seen these optical illusions—visuals interpretable in multiple ways.[1] They have been the subject of many a viral sensation that has made its way around the world. Of course, such illusions aren't new. Artists such as Salvador Dalí made a career painting them.[2] One of his famous works portrays the portrait of a distinguished old man, but when looked at carefully, it reveals a couple and a sleeping dog.

Neuroscientists have dialed into why such images trick our eyes. The brain takes different inputs from the environment and creates a model of how things work together. In time, as the input-to-event relationship becomes stable, consistently netting the same outcomes, the model becomes the framework for future interpretation in that context.

But the brain's reliance on frameworks comes with a few limitations. It overestimates certain factors, especially those that are of value to life and limb—eyes and faces that help us in our complex social lives; colors, such as red, that indicate blood and danger; and even certain sound frequencies and tones.[3] This is why the old man's face in the Dalí portrait jumps right out, while the couple in it isn't obvious unless you are told about it. But as soon as you are shown the hidden image, the brain now refines the old framework and applies the new one when the context reappears. That's why once an optical illusion, even a magic trick, is revealed, you recognize the sleight the next time it appears. You cannot unsee it.[4]

This recognition requires insight, someone or some event to unravel the illusion. Until then, the brain continues to apply whatever framework it possesses, even if it doesn't fit that context. The problem is that when something you apply doesn't suffice—say, you don't see the couple in the picture—you cannot know it is a suboptimal framework that doesn't work and refine it. But this refinement only occurs when someone points out the flaw in your framework, as in literally showing you the couple in the Dalí portrait.

This is difficult if everyone around you possesses the same flawed framework. It then becomes a bias everyone shares, which is how cultural and societal quirks emerge. For instance, in parts of India, people nod their heads from left to right to show agreement, something that throws off people used to nodding this way only to express disagreement. But it's a reaction that others in the group also follow. No one can point out the quirk, which then becomes the norm. Thus, having a valid and optimal framework is essential for thriving in today's complex social, cultural, and organizational contexts.

Recognizing this, human society has evolved a mechanism for developing valid frameworks. It is the activity of science—the purpose of which is to develop optimal frameworks for making sense of reality. The sciences, through investigation and repeated testing, develop models that best describe or predict individual and collective phenomena, which are then disseminated through our education system. The physical sciences focus on natural phenomena; the social sciences focus on human processes. Altogether, they provide a robust framework for interpreting and predicting people and processes in various contexts. Without them, much of what we witness would be interpreted based on our individual or collective biases. Our interpretation would be no different from that of a naive viewer looking at the Dalí portrait. We may get it right at times, but often we would miss what is hidden in plain view.

But even as science provides the robust frameworks, the complexity of human processes has led science to become highly specialized. This has led to hundreds of scientific subdisciplines—which at the turn of the nineteenth century were just a mere few areas of interest—each with their own journals, citation requirements, and worldviews. These have created strict disciplinary boundaries that scientists seldom cross and made it easier for them to become locked in collective quirks or paradigms of doing and thinking about things a certain way within a discipline.

In chapter 3, I discussed the engineering paradigm that dominates the security field's view of users. While most security engineers nod in agreement when discussing the need to study users, most apply frameworks from human factors research, regardless of their appropriateness. They treat computer users like industrial machine operators were treated at the turn of the twentieth century and focus on improving their security behaviors through repeated awareness training. Never has the field applied a framework from the social and cognitive sciences—especially one developed to explain user thought and action that leads to deception via social engineering or its detection. This has stymied how CISOs like Marks have dealt with users.

To appreciate this, let's look at the following 12 statements provided by NAA employees, which explain why they either did or didn't fall victim to the phishing pen test I'd developed:

1. "I quickly glanced at the email and it looked right."
2. "The colors and graphics were similar to our internal file sharing system."

3. "I thought I knew who the sender was in our company and so I focused on just a few graphics on the page to confirm it."
4. "I opened it on my phone, knowing it would be safer."
5. "I saw a PDF and opened it."
6. "I didn't download anything. What's the big deal with opening a link?"
7. "I didn't even think about what I was doing."
8. "The email came in just as I parked my car and I cannot stand unread email counts."
9. "I saw the dollar amount charged to my account and just reacted."
10. "I read and noticed the typos and bad grammar in the email."
11. "Of course it was fake—I remember someone telling us never to open emails not personally addressed to us."
12. "I looked closely at the email and found nothing wrong with it."

Depending on your experience and understanding of human psychology and behavior, these responses might make sense or they might appear to follow no logical pattern. You might recognize the user as trying to be safe, using some convoluted logic (e.g., I opened it on my phone . . . ), or perhaps a pattern of thinking you wouldn't have considered (e.g., I can't stand unread email counts). You might also miss the fact that the user was trying to be safe (e.g., I saw a PDF and opened it) or using a well-reasoned thought process (e.g., the colors and graphics were similar . . . file sharing system), because in these dozen statements there is a lot more information—some less obvious than it appears. In fact, there is purpose, reasoning, and a rationale hidden in every one of them.

Like the illusory couple hidden in the Dalí portrait, the reasons why people fall prey to phishing also hide in plain sight. All you need is an appropriate framework that changes your perspective, and, once revealed, it will be hard to unsee them. The sections that follow provide this framework.

### HOW WE ESTIMATE RISK

We begin with risk—a large part of what our brains are designed to estimate. Our brains have evolved to forecast risk to avoid the possibility of injury or death. These forecasts are based on contextual stimuli that are analyzed using beliefs about risk formed through prior experience in similar circumstances. Risk beliefs are, therefore, the starting point for unraveling why users are vulnerable to social engineering.

In general, risk beliefs can be accurate estimates, overestimates, or underestimates of risk. The degree to which the estimation digresses is contingent on three factors. The first is perceived task complexity. Some tasks (e.g., running) are obviously simple, whereas others, such as flying a plane, are clearly complex. The problem usually lies between these two poles. These are tasks that might *appear* simple. This is usually the case when people have analogous experiences that they can apply to them. For instance, playing the piano seems simple because it is similar to typing on a keyboard, a reason why the piano is among the most popular musical instruments people take to learning. At the other pole are tasks that *appear* complex. Many of the skills we learn in life, from bicycling to learning math, fall into this grouping. They appear complex because there is no ready referent in the mind that we can apply to them.

Such appearances cloud our estimation of risk, the effort the task will take, and the rewards from it. For instance, many people underestimate how much effort learning the piano actually entails, which, as any piano instructor will tell you, is the reason why most give up on it. At the other end, most students in the US begin with the belief that math is difficult, which clouds their experience with the subject and their willingness to pursue careers that require advanced knowledge of mathematics.

The second factor that influences the accuracy of risk beliefs is expertise. This is the degree to which the person has mastery over the performance of a task. Expertise can come from formal education and training and can also be acquired either vicariously or by asking others for guidance. To a good degree, expertise is contingent on the task's complexity. Simple tasks can be learned through observation and a little guidance. Running, cooking, bicycling, and many sports are examples. Harder tasks require advanced training and practice. You surely can't learn to fly a plane or scuba dive without instruction and practice.

The final factor, one that is often ignored, is metacognition. This is the self-knowledge of one's lack. It is the ability to know that what you know isn't sufficient. Metacognition distinguishes between those who, when faced with a task, seek information and those who do not. The latter individuals don't recognize that what they know doesn't suffice. Hence, they might rush into action and perform risky behaviors that could put their lives in jeopardy.

An example of this is the case of McArthur Wheeler, who robbed two banks in Pittsburgh, all while smiling at the security cameras in the banks that were pointing toward him. When arrested later, a puzzled Wheeler asked how the police had identified him. He had applied lemon juice on his face, which according to him should have rendered him invisible to the camera system, just as it makes ink invisible.

Wheeler's error in estimating risk spurred Cornell University psychologist David Dunning and his graduate student Justin Kruger to investigate the basis for such behaviors. Their finding, eponymously termed the Dunning-Kruger effect, implicates the lack of metacognition as causing an illusion of superiority. This minimizes the perceptions of risk in such people, because of which they perform riskier actions than those who have a more accurate view of their capabilities. Subsequent research on the Dunning-Kruger effect has found that such underestimation of risk is more common for low-skilled tasks, particularly those that appear easier to master.

As a case in point, take swimming. People in developed nations usually learn to swim through lessons in confined swimming pools. But it's easy to underestimate swimming's risks when outside the predictable boundaries of a pool, in open water, where according to the Centers for Disease Control and Prevention over 50 percent of fatal and nonfatal drowning accidents involving people 15 years of age or older occur. Many of these accidents likely occur because swimmers underestimate the risk and overestimate their own capabilities.

Or take bicycling, a relatively easy-to-master skill. Bicycling accidents account for the highest number of head injuries treated in hospital emergency rooms in the US. The vast majority of injuries are among adult males age 45 and over who are proficient at bicycling but weren't wearing a helmet. Here, too, we have a skill that appears deceptively easy and predictable, lulling individuals, especially those with low metacognition, into taking risky actions.

The same is the case with the internet. Many online actions that users perform and that social engineering attacks target—email, messaging, browsing, opening attachments and hyperlinks, using thumb drives—are deceptively easy. They appear so in part because their complexity has been deliberately simplified to foster adoption and use. This has been done through the development of graphical user interfaces (GUIs) with menus and user-friendly prompts (discussed in chapter 2) that have dumbed down the intricate choreography between machine language, operating systems, software, and hardware. Because of this, most users think they understand computing technologies and are skilled at using them. This disarms them and reduces their perceptions of risk.

Many online actions also appear simple to understand because of software design movements such as skewmorphism, as in programming things to appear like their real-world analogues. For example, email icons mimic the sights and sounds of mailing a postcard with symbols of paper clips and postcards added to strengthen this connection; digital books appear, even behave, like real-world books. These are programmed to foster a cognitive connection between the digital and the real, which below the visual surface are completely different. Similar ideas are fostered by terms like *debugging* and *antivirus* (which we discussed in chapter 3) that have nothing to do with their real-world counterparts. This layers bad ideas on top of bad ideas, routinizing the use of cognitive leaps to define underlying problems, which leads to fundamental errors in understanding online technologies and judging their associated risks.

McArthur Wheeler's behaviors were based on similar connections. He thought he understood how ink made from lemon juice disappears on paper, having likely seen it on a television program (given that it was 1995), but he also made a cognitive leap from the invisibility of ink on paper seen on camera to the invisibility of his face when seen on camera—connecting two completely unrelated factors. This is no different from what users do when they make connections between their inability to write on PDFs and the inability for malware to be injected into such files.

Wheeler's lowered risk perceptions disinhibited him, so not once did he stop to think through his actions. They short-circuited the cognitive processes that could have led to effective decision-making. In the case of Wheeler, they influenced his willingness to exert any cognitive effort in thinking through or testing the disappearance powers of lemon juice. In the online case, lower cyber risk belief (CRB) estimates likewise influence users' willingness to exert cognitive effort into examining the details of, say, a PDF document in an email, or any email that is opened on a perceived safe network, device, system, or software.

Just like testing invisible ink's effectiveness on a surveillance camera system was outside the range of Wheeler's capabilities, it is hard, if not impossible, for the average computer user to inject malware into PDFs or different operating systems and assess their relative resilience. Consequently, users with CRBs who underestimate risk tend to be more cavalier in their online action. They are more likely to open emails, even ones they find questionable, on work computers, on certain operating systems, or while using certain types of wireline or wireless connections, once they believe they are less risky.

If you conduct pen tests, you may have wondered why some users, even after being told repeatedly not to open questionable websites or not to reuse their passwords, continue to do so, or why, no matter the training, users continue to open the phishing pen-test email. You now know why. Users underestimate the cyber risk from such actions. Thus, CRB reveals what users, when faced by a phishing attack, might choose to evaluate in the email, how much cognitive effort they might expend, and, finally, how they think about the attack.

But what about confidence or technical proficiency? Don't people fall for phishing simply because they are overconfident and lack technical skills? Isn't that why we should scare them through phishing pen tests and train them repeatedly?

Confidence and technical proficiency are surface-level explanations. They work on a post hoc basis for blaming people. For instance, whenever an adult who knows how to swim accidentally drowns in open water, it is easy to blame their overconfidence or their lack of "sufficient" proficiency, but it is hard to explain how much of their confidence was "over" the threshold beyond necessary or what level of swimming proficiency would have been sufficient. Simply put, they provide convenient answers, not actionable solutions.

The same is the case with technology users. There is no accurate, objective metric of security confidence or technical proficiency, so there is no way to tell when someone's security confidence is too much, let alone when their technical proficiency is sufficient. Because of this, there is no way to train users to meet these requirements. In contrast to this, CRBs explain the specific mind-level activities that lead to an overestimation of risk, a perception of mastery, or a false sense of competency. It explains rather than castigates and, more importantly, pinpoints the specific beliefs that lead to the estimations, so we can develop solutions that help fix the problem at its source—in the mind of the user.

### HOW WE THINK

People all over the world eat breakfast, lunch, and dinner; there are snacks in between and fluids consumed throughout the day. One-third of this is to feed the energy requirements of our brain, which is a mere 2–3 percent of an adult human's mass. These energy requirements increase further when this tiny organ is actively using its capabilities for thinking and responding to stimuli. We have therefore evolved strategies for achieving efficiency in how it deploys its resources.[5]

One strategy involves adapting information and incorporating it into existing structures of knowledge. This happens right from our formative years. For instance, a flash card teaches children to associate the letter Z with an animal image of a zebra.

This connection is reinforced by exposure to photographs, videos, and visits to the zoo. The brain then forms a connection between the animal and details of its colors, shape, size, how it moves, what it eats, and where it lives. Repetition and continued exposure reinforce these connections. Once these connections are created, the child's brain no longer re-creates every visual detail about zebras each time its name or something reminiscent of the animal is evoked. It would be too energy intensive if it had to do this repeatedly. Instead, the brain forms a heuristic, a thumb rule, which connects salient aspects of the animal's colors, shape, look, and feel with the concept zebra, which the brain uses to conserve cognitive energy.[6]

Heuristics are in the form of "if, then" statements, such as "*if* a hoofed animal has black and white stripes, *then* it is likely a zebra." So, whenever children see a few cues—a look, a feel, a color, a shape—associated with a zebra, they can quickly arrive at the logical conclusion that it is likely a zebra.

We form many such heuristics throughout our lives, some simple, others more complex. For instance, the same black and white stripes that trigger a zebra are also connected to pedestrian crossings, aptly called "zebra crossings" in some countries. Many heuristics are learned during our formative years, others through schooling and lived experience. As we move through life, we might connect the word *zebra* to a type of freshwater fish, a style of adult clothing, a music band, and so on, layering and mixing cues of increasing complexity.

Three things are necessary for the brain to trigger any such heuristic during decision making. First, cues need to be present or *available* in the information space. In the case of zebra crossings, there needs to be something associated with black and white stripes present for the "*if* black and white stripes, *then* it is safe to cross" heuristic to be triggered. It could be a look, the presence of other cues (e.g., a traffic light), or even a description. Second, heuristics need to be *accessible* in the user's memory. That is, there needs to be a heuristic already formed in the person's memory linking such stripes on streets to the "if-then" decision; without it, the cues would be meaningless. Being *available* and *accessible* of course doesn't mean much unless the user is motivated to notice the cues.[7] Thus, the final requirement is that users perceive some risk, such as believing it is risky to cross a street, which makes them look for a safe path across.

Heuristics can be triggered by the mere hint of cues. They help us make snap decisions, which allows the brain to conserve cognitive energy. The importance of cues as quick signals of action has led human societies to evolve ways of standardizing them. We developed uniforms for police, pilots, and physicians; we designed emergency vehicle sirens, light colors, and symbols; and we developed traffic lights and zebra crossing marks, which we apply more or less consistently all over the world. Companies took it further and developed brand logos and trademarks, with accompanying sights, sounds, and colors. For instance, in Starbucks coffee shops around the world, the logo, colors, furniture, posters on the wall, lights, the store's ambience, and even the music are purposefully and painstakingly kept consistent. All this has one reason: to serve as a cue about what to expect from a Starbucks product. This way, consumers don't have to think much about the quality of the product they are about to experience no matter where they are.

The virtual space is similar to the offline space but with fewer sensory inputs. Most of us seldom receive unique auditory signals online, at least not since AOL's "You Got Mail!" spoken email chime went out of vogue. Olfactory or smell-based signals are nonexistent, while kinesthetic signals are limited to haptic feedback through touch pads and incoming message indicators that are undifferentiated and lack nuance. Online differentiation primarily relies on visual cues, much of it involving GUI buttons or graphical icons that signal action. (We discussed in chapter 2 how GUI developed and altered the computing experience.) Today, be it through mouse clicks or touch, whatever users do on their computing devices is done through their interaction with such icons.

Icons trigger programs to open; icons within them initiate drop-down menus with even more icons. They also serve as a signal of a product or experience. These include singular icons such as the SSL lock icon next to a website's name on browsers, which is used to signify that the website's traffic is encrypted. Different icons work in tandem with colors, fonts, text, layouts, form fields, and graphics that are organized in a consistent manner. Just as Starbucks does within its stores, online icons are arranged to differentiate between spaces and foster a heuristic about what the user is about to experience when they click on different cues. For instance, the Facebook logo, logo color, page color, font, and page layout have been consistently presented, so users now associate them with the brand's social media experience. Each has associated heuristics about the Facebook experience. This is why when users are asked to log in using their social media credentials on a different website (such as to comment on a news story or log onto another service), all they need to see are a few cues from the Facebook brand to enter their Facebook credentials. A completely different set of colors, layout, fonts, and logos are associated with the LinkedIn site and triggers a distinct set of heuristics related to that experience.

But here's where things get even more complicated. In the real world, the cue to a heuristic relationship can be scrutinized and confirmed. You can find out for certain whether a horse is different from a zebra. Their associated cues and heuristics can be experienced, verified, and refined. Much of computing, however, utilizes visual cues that are built on top of more cues. Each is a virtual construct that has no basis in reality.

Take the SSL lock icon mentioned earlier, which many in security circles check to identify whether a website's traffic is encrypted. Let's assume that the symbol for it is *available, accessible,* and that users are motivated, as in having cyber risk beliefs that drive them to look for it on browsers. Now assessing the SSL is the virtual equivalent of checking whether a zebra is indeed one and not just a horse painted with stripes. Checking whether a website's traffic is encrypted involves seeing whether the website's address begins with https. Further assessment involves clicking on the lock icon next to the site's name in a browser. Doing so opens a window that tells you about the type of certificate and the certificate authority (CA) that issued it, with hyperlinks that lead to the CA's website. Go further and you are led to more web pages that present information about the CA. Each step leads to a construct, a virtual concept we have all agreed on: what http means, what the addition of "s" to it stands for, what a certificate is, who the CA is and how they certify a website, who allows them to do this, for how long, and on and on. Each is a series of cues that lead to others, with none of the cues used in the certificate assessment process providing absolute certainty that a website is indeed encrypted.

But this isn't the only issue. The cue-to-heuristic relationship in virtual spaces is further complicated by the nature of cues in computing in a number of different ways. For one thing, the online environment is rife with cues. Not only is there an SSL cue on the browser, but there are dozens more on the browser, the website the browser brings up, on other programs on the device, and on the device itself. Many are available in the same visual plane, meaning they appear simultaneously, each vying for the user's attention.

Second, while many cues exist, some are more obvious than others because of their prominence and their relevance within the decision context. For instance, as soon as Google's search website opens up, the search bar form field becomes prominent and the user's focus immediately turns to it, but when the search results are returned, the individual search pages, promoted hyperlinks, and images become central. Against all these prominently placed and highly visible cues, the tiny lock indicating SSL next to the domain name competes for attention.

Third, many cues are implicit. This is both a consequence of the evolution of interfaces and design choices made by programmers. Take any email program from shortly after 2000 and you will notice that even simple Reply or Forward icons were labeled and spelled out. Now check out the Yahoo web portal of today and on most browsers. Yahoo's programmers provide small user interface icons, such as a curved arrow on the bottom of the screen, that they presume users recognize. This move toward making icons implicit is even more pronounced on mobile apps, where the screen sizes are limited and more information needs to fit into the visual space. Because of this, users form heuristics based on subtle cues that they use to intuitively navigate websites. This makes all manner of cues, from the most obvious to the subtle, capable of triggering heuristics.

Fourth, the placement of cues, even the same cues, at times varies considerably. For instance, every browser has a different location for displaying the SSL symbol. Even the same browser permits different actions on different instances of use. For instance, Google's Chrome browser allows users to assess the quality of the SSL certificate by touching or clicking the lock icon on computers and mobile devices. In contrast, Apple's Safari browser allows the SSL icon to be clicked on and examined on a computer but not on iOS devices.

The fifth factor complicating the cue-to-heuristic relationship is how cues are placed. During the user interface development process, programmers make explicit choices about which GUI elements are most necessary. In almost every case, these tend to be action buttons that are at the core of the product's performance. Email apps, be they on their web portals or their mobile apps, prioritize opening and responding to email; social media apps prioritize accepting connection requests; and browsers prioritize opening and navigating websites rather than assessing their SSL certificates. Most seldom centralize delaying a response, verifying a sender, looking at source codes, assessing a sender's IP address, or examining the quality of the SSL or its CA.

The sixth complicating factor is that even the absence of a cue can trigger heuristics. Take the case of the SSL icon. Most security training programs train users to look for the icon and to never enter sensitive information on a website that doesn't have it, so the absence of the SSL icon has an associated heuristic just as much as its presence triggers.

The seventh factor influencing the cue-to-heuristic relationship is that there is no gatekeeper or higher-order credible authority who can with reasonable certainty vouch for the authenticity of an email address, website, online service, patch, or update. Online veracity can only be assessed by clicking on cues that lead to more cues, which lead to even more cues.

The final factor complicating the GUI-cue-heuristic relationship is that much of what is done online has little to no direct consequence on the life of the organism—the user. Unlike the cue to a "zebra crossing," missing which might lead to a potential loss of life, the consequence of opening a nonencrypted page appears minimal and is indirect. The human brain is therefore more attentive to signs of danger in the real world. In contrast, users' cyber risk perceptions lack both nuance and acuity because of the newness of the virtual environment and also because the need to evolve these capabilities doesn't appear to be pressing.

Thus, we have several issues with virtual cues, their presentation, their prominence, their placement, their positioning, their overall purpose, and their value to computer users. Users' brains have been conditioned to accommodate a variety of such cue-related issues while in virtual spaces in a matter of seconds. To accommodate them, most users rely on heuristics, at times the same heuristic across multiple conflicting cues. For instance, an Apple customer learns that the emails emanating from multiple conflicting email addresses are all legitimate communications with Apple. For example, communication from Apple's new credit card systems are emailed from an "@post.applecard.apple" domain, while its iTunes music store receipts come from an "@itunes.com" mailbox. Some of these emails have attached documents, others hyperlink to external sites, and still others have image files. Likewise, users sending and receiving files using any cloud sharing service quickly learn that the hyperlinks, even email addresses used for sharing, have a wide range of symbols and characters. For instance, Google Drive notifications come from a "drive-share-noreply@google.com" in-box and Dropbox notifications come from a "no-reply@dropbox.com" in-box. Sometimes, as in the Google Photos example, the name of the person sharing a photograph via email through their sharing feature appears and remains the same. Every sharing email, however, comes from "noreply -010203c023b2d094394a@google.com," where the alphanumeric characters (randomly chosen for this example) change each time.

These examples aren't exceptions. Because of the variety of add-on services that companies like Google, Apple, Amazon, and others provide, they are endemic to the online user experience. In all such instances, users' brains have to keep accommodating highly nuanced variations. Rather than do this, most who don't believe there is much cyber risk are motivated to minimize cognitive effort. They resort to using broadly applicable heuristics. "*If* Apple/Google/Amazon or whoever uses other email addresses, *then* it is authentic."

They seldom stop to consider whether such heuristics are appropriate or sufficient. Think of the last time you stopped after typing out a text message and postponed sending it, typed out a website's name but did not go onto it, or abandoned a search after typing the search phrase. Most of us seldom do these things because the design of these functions fosters moving ahead. They are like the escalators in airports and gangways in sports arenas, designed to move you toward your eventual destination. In this case, it's doing whatever the service was designed to do, not stop along the way and evaluate the sights. Online interface designs not only present cues and trigger a variety of heuristics, they foster action on them.

This makes it easy for social engineers to deceive users. All they need to do is inject a few salient cues from certain services or reproduce a few keywords to trigger associated heuristics. They need to do just enough to trigger the heuristic.

For instance, they could create a phishing email using Amazon's name, font, colors, and layout and easily trick users into believing it was credible. They can mimic cloud-based file-sharing links from a variety of in-boxes without caring about the sender's email address, because they know users accommodate, even expect, such variations, or they can simply inject a form field onto a login page using, say, the colors of Gmail and find users willingly entering their credentials. The rest of it, the site or the app, which fostered the cue-to-heuristic association and made it available and accessible, ensures this because it moves you forward to the decision by making you click, open, download, or enter credentials. And that's how you fall victim to spear phishing.

But there is another form of processing every one of us can use. Cognitive science research has shown that the human mind is capable of performing two forms of information processing. This second mode is more focused on deriving meaning from information in the context. It connects information to its relevance, to what it means, even what is implied by it. It is elaborate, purposeful, and careful. It is called *systematic processing*.

In systematic processing, our brain actively connects semantic information from within the information context to prior knowledge held in memory. Given its focused nature, such processing requires that users be motivated to expend cognitive energy. Additionally, the outcome of systematic processing is predicated on the quality of prior knowledge, without which the outcome of the processing effort would remain suboptimal.

Early experimental research on spear phishing detection found that users who systematically processed emails were significantly more likely to detect the deception.[8] This finding inspired many of today's cybersecurity training protocols that try to motivate systematic processing and improve users' domain knowledge, but real-world pen tests and ecologically valid experiments (as in those that re-create real-world email use) have found paradoxical results, with users seldom engaging in systematic processing of emails or applying their knowledge. The reason for this is that the early experiments were creating contrived situations where users were shown multiple emails and asked to spot the fraudulent ones. It is no different from asking people to go on a treasure hunt. Knowing there is a treasure to hunt increases the focus on interpreting clues. This very act motivates care in processing and the deployment of cognitive resources toward it, which enhances the chance of deception detection.

But, as discussed earlier, in everyday computing, users are presented with a myriad of cues in emails, messages, and communications. Compared to text, cues are prominent, graphical, and colorful. They have associated heuristics, each signaling movement and action (send, reply, edit, and so on). Compare this with reading text that varies and for which substantive interpretation requires focused reading and comprehending. Extracting meaning requires the deployment of cognitive resources and, depending on the context, sometimes in significant amounts, which the brain is designed to do sparingly. This is why systematic processing occurs sporadically and for short periods of time. Even when it does occur, systematic processing efforts are made after heuristics, triggered by cues in the context that tend to be more prominent and easy to interpret, have biased the interpretive task.

For instance, when encountering an unfamiliar email from, say, Apple, it is easy to see how an Apple logo in the email, which is usually larger and of a different font and color than all the surrounding context, might bias the subsequent systematic evaluation of the domain name from which the email was sent. An unfamiliar domain name might even be discounted by the "*if* it has the Apple logo or even just the name Apple in it, *then* it could be from one of their many service in-boxes" heuristic.

Furthermore, systematically scrutinizing the originating IP or any other detailed information beyond the domain name might be difficult because some programs discourage it. In fact, Apple's iOS Mail does not even display the source header, so accomplishing this on an iPhone requires that the user switch to a different device or another mail client. Most users seldom do all this. The vast majority are conditioned by the conveyor belt style design of websites and apps to complete whatever online interaction they have begun on the device or app in front of them, not to evaluate, postpone, read, or revise. They finish what they have begun. They don't think it is risky or that what they are about to do will cause them harm. Their cyber risk beliefs don't motivate systematic scrutiny. Thus, even though systematic processing in theory increases deception detection, users are seldom willing or able to engage in it.

Social engineers understand this. They realize that users are miserly in their cognitive effort and that they are conditioned to react to cues and make snap judgments. That is why they focus on graphical cues that mimic online brands, apps, services, and actions. It is also why they seldom worry about textual and grammatical errors in such attacks—because they know that users seldom carefully read through them.

In many ways, hackers are like cognitive scientists. They experiment with different attacks to assess which ones work best, they test different cue combinations to identify the cues that matter, and they work to improve their attacks by constantly studying users' reactions.

Understanding how the human brain usually processes information tells us a lot about why social engineering succeeds and why people fall victim to it, but understanding how the brain functions under unusual circumstances tells us even more.

### DOING WITHOUT THINKING

A patient identified in the medical literature by the initials DF suffered extensive brain damage caused by carbon monoxide poisoning. It caused a unique form of visual agnosia, where she could see colors and visual textures but couldn't identify even the simplest of geometric figures by shape. Yet, when she reached out to pick up such objects, the spacing between her thumb and finger matched their ideal grasp points. She could do the same with smooth objects. She could also accurately orient her wrists to mail a letter through a slot—all tasks that require visual acuity.[9]

In another case, viral encephalitis destroyed Eugene Pauly's brain cells such that it was impossible for him to remember things he had done mere minutes ago, but he could perform activities such as taking a regular morning walk around the neighborhood and find his way back home. He could even put on the TV, surf channels, and find his favorite TV program. He could somehow learn new behaviors, just as long as they were repeated and followed patterned sequences.[10]

Patients such as DF and Pauly prompted a new understanding of the disassociation between thoughts and action. It appears we have two senses, one for thought-based actions and another for automatic actions. Actions that are frequently repeated are stored in a primitive part of our brain as habits, separate from where immediate memory that is used during cognitive processing is retained. This gives us the ability to perform some of our repeated sequential behaviors without conscious awareness.

We needn't look far for evidence of this because every one of us engages in many nonconscious habits throughout the day. Most of us cannot recall in which direction we brushed our teeth or, if we regularly drive to work, the number of traffic signals we stopped at along the way, at least not until we are specifically asked to recall them. Many of us might not recall when we switched on a light or how we oriented our palm when we last did so, and if we use multiple devices to check different email accounts, which device or account we last checked. We can form habits for simple repeated actions, such as putting on the TV at a certain time of the day, and even for complex actions, such as operating a car. Some can be benign actions, such as mindlessly checking social media feeds. Others can border on the risky, such as driving or crossing the road while texting.

Most of these begin as conscious reactions to specific cues for achieving certain gratifying goals. These could be environmental cues such as a clock indicating the time of the day cueing us to put on the coffee pot, or contextual cues such as the availability of a phone or an app for checking our social media feeds to see what our friends are up to. Over time, the sequence of actions required to perform the behavior to receive the gratification becomes predictable. Our brain consequently stores the sequence separately as an action script that is then applied whenever the cue is present, such as whenever the clock shows a certain time. Action scripts create cognitive efficiency because the brain doesn't have to think before it acts. This makes such scripts ideal for reacting quickly.

For users who receive some kind of pleasure—enjoyment, social approval, escapism[11]—by doing some action, the cue-action-gratification relationship can get so blurred that strong habits are formed, where the actions could be triggered by the urge to achieve gratification even in the cue's absence. For instance, for users who find approval in social media and messaging, the habitual urge to check their media feeds, usually cued whenever notifications show up, can become so strong that they may keep reaching for their device to look for notifications and updates. This is in contrast to utilitarian or functional outcomes that are sought from repetitive behaviors (e.g., brushing teeth, turning on a switch). Their underlying motivation is efficiency and expediency, which lead to weaker habits, because these behaviors are examined more often and corrected. Take the case of safely crossing the road by quickly glancing on the left side for oncoming traffic—an action script I developed while living in the US. This habit failed me whenever I traveled to countries such as Singapore, where oncoming traffic comes from the right side of the road. After a few close calls, I consciously revised the pattern by replacing it with looking at both sides of the street before crossing regardless of where I travel.

Among modern computing devices, the frequent use of smartphones and portable computing devices in particular leads to strong habits. This is partly because of their design and partly happenstance.

Steve Jobs gets the design blame. In the original iPhone, he deliberately excluded an incoming message-notifying LED, which was then common in smartphone brands such as Blackberry. The LED's colors and the frequency of flashes could be programmed such that users only needed to interact with their phones when they lit up. Jobs's design decision, which was followed by all leading smartphone makers, made it necessary for users to pick up their phones and check for incoming mail and messages on their screens. This created a new pattern of checking one's phone, the frequency of which increased as more apps were developed that started sending notifications.

Then came happenstance. With the success of the iPhone and the rise of consumer-focused email services, what used to be business-focused "push" email services and messaging applications that sent email to devices when they arrived, got replaced by email that was "pulled" from servers and delivered at intervals defined by the device or the email's provider. This made the outcome of checking smartphones unpredictable. It fostered constant checking for the occasional receipt of a message.

The next addition to the mix was an app for almost every service, which previously had been limited to browser-based interactions. From shopping to auctions, gaming, and photo sharing, there were now many more gratifying reasons for checking one's phone. Thanks to them, phones became more than just communication devices. They became the conduit for enjoyment, fantasy, and escapism. Many were taking on social characteristics—presenting user reviews, allowing user reports, and pushing notifications to user devices.

The last addition was mobile-only services, such as ride sharing, gaming, and payment systems, that were designed solely for smartphones and communicated via smartphone notifications, emails, and texts. Checking a smartphone frequently went from being controlled to fun to a necessity.

Thanks to this, notifications that triggered a reactive response accompanied a variety of apps, messages, and updates. There were apps providing enjoyment, social approval, and escapism, and notifications were connecting users to them. Altogether, they created a Pavlovian learning process with constant checking in response to varying sounds and symbols, leading to an occasional gratifying return—a relevant message, email, news story, or status update. This created a variable reinforcement loop no different from the system behind slot machines in casinos that lead to a gambling habit, only it was on an always-on smartphone that users carried all the time.

In time, the process led users to reactively check their phones and even diminished the need for notifications, making picking up one's phone simply for the sake of checking for messages a common habit. This smartphone habit became so widespread that it caused major accidents and public safety problems. Some cities started ticketing people for walking or driving while checking their devices. Even smartphone makers acknowledged the problem and developed apps to limit device use.

But such habits did not simply end with checking phones. They extended to how users responded to notifications on them. With the bulk of notifications stemming from email, messaging applications, and social media, users were reacting to requests on these applications without thinking much about the request. One reason was that the notifications were far too many, at least in the early days of apps, when systems to limit notifications were not yet in place. The other reason was that most requests, although gratifying, were routine—like something, rate something, upvote something—and mostly inconsequential. Yet another reason was that users were doing other things simultaneously when checking these apps, especially while on mobile devices. They were usually walking, talking, driving, using other apps, or swiping away at notifications from other apps, all in the relatively constrained visual plane of a smartphone. This made it even harder for users to pay attention.

Take the story narrated by the former head of the FBI, Robert Muller, at the Commonwealth Club of California in 2009. Muller confessed that he blindly followed the instructions in a phishing email and typed in his banking information, but a mere click or so away from falling for the attack, he "caught himself in time" and realized that it "might not be such a good idea."[12] Muller didn't specify the device he was on. Let's presume he was on a computer, using a full browser, and looking solely at this email.

But picture the same request on a smartphone. You have a smaller screen and a mobile operating system. It presents a mobile-friendly visual of the page that limits what you see. The browser further restricts what you can access (such as the SSL indicators) and your ability to navigate to other pages to check whether the request is authentic. Now imagine looking at this request while you are walking or talking or, worse yet, driving. Add some notifications to the mix that are frequently interrupting you, vying for your attention. Now consider how mindful you would be in your thinking.

And it's not just catching yourself in midthought. It is about stopping yourself from doing something reactively on these devices in mere seconds: opening hyperlinks, opening attachments, accepting social media requests, enabling macros in a program, entering passwords, or responding to messages. These are the actions that users must stop.

The challenge is that smartphones and their apps are deliberately designed to make users react. For instance, the "Accept" icon is centrally placed on social media friend requests, email apps highlight active hyperlinks on cloud sharing notifications, the send icon is prominent in email programs, and the icons for accepting phone calls are larger and more prominent than those for blocking them. Login and password entry fields and search bars are likewise prominent, as are buttons for confirming, entering, and agreeing to requests. These are deliberately programmed to require little effort from the user—touch the screen, enter a password, perform a few swipe gestures, accept a request, or simply look at the screen. Repeated reactions to them are egged on by constantly updating email counts, message indicators, last seen markers, update notices, and incoming notifications. Because of this, users are routinized into opening, checking, and accepting requests on their smartphones. Some users have even developed quirky rituals, such as not wanting to see an unread email count or notification that they have not swiped away. With computing operating systems and programs increasingly emulating their mobile counterparts, these quirks and patterns manifest themselves in general computing environments as well. Thus, the design of computing fosters habits that cause a reactive acceptance of social engineering requests on them.

But not everyone develops such habits. There are certain personality factors that make some more prone to it. One factor appears to be conscientiousness. It is the degree to which the person can exercise self-discipline over their actions. The other factor appears to be emotional stability, or how well a person can control their impulses. Together, they define the extent to which thoughts can induce a reaction. They dictate whether one person will, say, view the unread email count on their device and decide to ignore it, while another might feel compelled to check the messages. They tell us which user will likely react adversely to a threat or warning in an email and who might choose to respond reasonably. Individuals with lower levels of emotional stability and conscientiousness thus perform actions impulsively. They are termed *deficient in self-regulation* (DSR).[13]

Research has shown DSR to be a marker of users' likelihood of developing strong media habits. It works in tandem with the gratifications they seek. For users with DSR who seek enjoyment, approval, and escapism from their media use, the gratifications are like a carrot that they cannot stop trying to grasp. Their impulsive actions become compulsive because of the design of phones and apps, which foster repeated reactions. The trifecta of influences—DSR, personal gratifications obtained, and device design—creates behavioral loops that are increasingly impulsive and lack reasoning, which, over time, leads to stronger media habits among such users.

Strong habits aren't necessarily bad. They are just mindlessly performed, which makes it possible for hackers to co-opt them and make users do things online that could lead to a breach. They can become bad habits.

In experiments I have conducted, those with strong social media habits are more likely to provide personal information in response to a social engineering request that appears like a routine request. This includes accepting phony social media friend requests, complying with internal requests for information, and clicking on hyperlinks and attachments sent to them via social media. The same trend can be found among victims of email-based phishing. Users with strong email habits are far more likely to click on a phishing link within an email or open a malicious attachment than users with weaker habits.

Thus, habits are an important set of impulses that directly influence user vulnerability to social engineering. Their influences range from making people rush to judgment by invoking a reactive response to, depending on the habit's strength, altogether eliminating any thought prior to clicking on a hyperlink or a malicious attachment in a phishing email.

### INTERPRETING THE RESPONSES OF NATIONAL ACCOUNTING AGENCY (NAA) USERS

Let's recap what we have discussed so far in this chapter. Cognitive heuristics are intuitive decisions triggered by cues; they are held in the form of "if, then" rules. Systematic processing requires making semantic connections between elements in the information context with knowledge. Both are premised by the users' beliefs about the risks of online actions: cyber risk beliefs. These are the major cognitive factors defining how users respond to social engineering. We also have habitual reactions, which are premised on personality-based regulatory abilities (as in deficient self-regulation) that lead to automatic, reactive responses to social engineering.

Now let's revisit the responses provided by the NAA's users explaining why they fell victim to the phishing test attack. Examine each statement using the framework.

The first response, "I quickly glanced at the page and it looked right," is from a user who connected the "look" of the email to its authenticity. The user had a mental rule that was applied: "*if* it appears similar to what I usually see, *then* it must be authentic." Recall that the email I crafted borrowed visual elements such as the color, font types, and layout from the official file-sharing system. The user's heuristic came from a quick glance of the email and its fit with their mental schema of how such emails usually appear. Thus, this user applied a cognitive heuristic.

The second response, "The colors and graphics were similar to our internal file sharing system," reflects even more explicitly the heuristic that "*if* colors and graphics are similar, *then* it must be real."

The third user looked at the source of the email and used that as a heuristic to guide the subsequent scrutiny. The user's "*if* it is from someone I know, *then* it must likely be authentic" heuristic biased the rest of the assessment.

The next user stated, "I opened it on my iPhone, knowing it would be safer." This user clearly invoked a belief about the safety of their device's operating system, a cyber risk belief.

The same is obvious in the next two responses from users, each of whom had a belief about the inherent risk of the technologies or actions that guided their decisions. One saw a PDF—and we know how users think of such document types—and the other thought clicking on a link wasn't as risky as downloading something. So far, the responses from these users reveal either a thought triggered by some element in the email or a preexisting belief.

The next set of user statements suggests less-mindful decisions. One user stated, "I didn't even think about what I was doing"; another stated, "The email came in on my phone just as I had parked my car," revealing the constraint in thinking they faced; a third user explicitly stated how they "saw the dollar amount charged to my account" and reacted. Automaticity, lack of thought, impulsiveness, and reactivity, all hallmarks of habit, are apparent.

The final set of statements is different. The first respondent read the email carefully and noticed typos and grammatical errors. It doesn't take much to guess that this person was thinking carefully.

The same is obvious from the next user, who, like the previous user, methodically thought through the email and connected it to prior knowledge about such attacks. Neither of them fell victim to the attack, because they systematically processed it.

But what about the last user, the one who "looked closely at the email and found nothing wrong with it"? This user also deeply elaborated on the information in the email but fell for the attack. What's clear from the response is that the user didn't know what to look for in the email. Their response demonstrates how systematic processing alone does not insure against deception. It only helps when the person is equipped with accurate knowledge.

Thus, each response reveals a lot about the user. Each provides insights into the user's beliefs, thoughts, ways of thinking, and actions, together explaining why the user was either deceived or detected the deception. In other words, they help diagnose what's really going on in the user's mind.

But such responses aren't hard to gather. In fact, we have been gathering such sentiments all along. For instance, users in the awareness training research conducted by MITRE that I discussed in chapter 3 and in many other research studies have provided open-ended responses, explaining why they fell for a spear phishing pen test.[14] The issue was that these responses were never examined using a theoretical framework based on social or cognitive science. The responses were given cursory attention and, like the archaeological findings of fossilized bones in ancient Greek ruins we discussed in chapter 3, were discarded as accidental.

While ignoring the obvious, the field of cybersecurity instead focused on training users to do things they are cognitively incapable of doing—systematically examining emails for prolonged periods of time; checking dozens of elements in each and every email; or being vigilant regarding attacks on email, messaging, or social media. Users were expected to perform these tasks on devices, apps, and operating systems that were designed for quick thinking, multitasking, habitual checking, and reactive responding, and when the users remained vulnerable to social engineering, they were retrained, retested, cautioned, and then blamed for their failure.

But now you know better. You know what drives users' thoughts and actions that lead to social engineering deception. You know how they think, why they think the way they do, and also when they are likely to act in ways they shouldn't. You know how online cues trigger users' thoughts, how devices influence their behavior, and how cyber risk beliefs motivate them. Chapter 5 goes even further, providing a simple mechanism for quantitatively pinpointing the most vulnerable users—the weakest links. Armed with this knowledge, instead of saying "Gotcha," IT managers can start saying to their users "I got you!"

# 5 THE KEY SYMPTOM

Most motor vehicle crashes in the US are caused by driver error. Mistakes made by drivers account for 96 percent of crashes at intersections and 92 percent of those on the open road.[1] With this in mind, take New York State Route 198, also called the Scajaquada Expressway, which runs through Buffalo, New York. If you drive along this expressway, chances are disproportionately high that you will witness an accident or, if you are unlucky, have one. That is because 50 percent of the expressway's open stretches and one in four of its intersections have accidents at rates higher than the average for any other similar road in New York State.[2] Could we conclude that Buffalo has the worst drivers in the state of New York? Of course not! A closer look at the Scajaquada Expressway explains why not.

First, this expressway has several S curves that are hard to navigate in the best of conditions, let alone in the snow and ice during Buffalo's infamously harsh winters. Second, when the Scajaquada was built in the 1960s, there was significant opposition from residents who disliked how it split different neighborhoods. To accommodate them, the expressway was hastily redesigned, which created some odd traffic patterns. For instance, when exiting one of its westbound intersections, drivers navigate four lanes of traffic coming north, south, and from another four-lane road abutting it. That's 12 lanes of traffic going in three different directions, not counting the entrance ramp and, on the other side, the exit ramp. On top of this, drivers in each direction have to contend with a variety of stop signs, yield signs, and traffic lights that are only partially visible to them because of the upward slope of the exits and the downward slope of lanes. All this makes driving on these roads confusing even for people who regularly ply them. Finally, the expressway circumscribes many of the area's cultural and recreational spots. This adds a nearby subway station, bus stops, pedestrians, and bicyclists to the already confusing mix. When you factor in weather, pedestrians, and poor planning, the accident rate shouldn't come as a surprise. Considering the circumstances, the drivers here have just as many if not fewer, accidents than others in the state.

Accidents are like a person's heart that stops pumping blood. This happens when everyone dies (unless they are already brain-dead), but it isn't the actual cause of death. It is a symptom of death, not a diagnosis for how not to die and live long. Likewise, an auto accident, even a lot of them, is merely a symptom. While driver error may lead to accidents on roads like the Scajaquada, the underlying reasons for it could be anything from internal distractions to inadequate surveillance and wrong assumptions about another driver's actions. It could also be precipitated by external factors outside the control of the user, such as weather, visibility, car design, speed limits, and poor street signage. A mere count of accidents doesn't justify blaming the user. Nor does it help in understanding why accidents happen or how we can avoid them. Yet, this is precisely what we are doing in cybersecurity today. Not only are we counting users' "accidents" after pen tests, but we are also using them to diagnose and reduce cyber risk.

Take, for instance, SANS Institute, the leading provider of cybersecurity training, which has certified over 165,000 security professionals all over the world.[3] SANS has developed a Security Awareness Roadmap, where it provides guidelines on how organizations should measure user cyber risk. It asks IT managers the world over to track one metric, what it calls an impact metric: the number of people who have failed a phishing pen test. This is equivalent to measuring the number of phishing accidents that occurred in the organization. There is not much else that is tracked by that metric—no reasons, no causes, no precipitating factors, just the accident rate.

Besides this, the SANS road map also asks IT managers to track what it calls compliance metrics. This includes the number of users who have completed training, the number of users who have signed the acceptable use policy, and the number of on-site training sessions they have completed. It advocates counting the number of sensitive documents in dumpsters, the number of mobile devices with passcodes, and the number of infected systems each month. It also includes two attitudinal questions, assessed using a survey, which measure the general attitudes of users toward information security and whether users believe their co-workers could have shared their passwords.

None of these metrics is of much value for diagnosing what led to the users to fail the phishing pen test. They either measure invariant behaviors (e.g., everyone has to sign the acceptable use policy, and that, too, just once) or behaviors outside the strict purview of the IT department (e.g., most smartphones require a passcode, and the IT department doesn't control those passcodes), and their relationship with user resilience and failures on phishing pen tests is questionable (e.g., Does putting sensitive documents in a dumpster reduce phishing failures? Do positive feelings toward cybersecurity or mistrust of co-workers translate to cybersafety?). If we were to use the analogy of motor vehicle drivers in Buffalo, New York, our compliance data would be the number of drivers who have procured a driver's license—which is everyone—and the impact metric would be whether they had an accident. Thus, the backbone of the SANS road map is the measurement of the rate of phishing accidents. It is the metric of cyber risk and the diagnosis of impact of the phishing pen test.

SANS isn't alone in advocating this metric. It developed the guideline by pooling the suggestions of security professionals actively involved in security awareness training, so the road map is a consensus document, reflecting the judgment of the cybersecurity professional community. In chapter 3, we discussed how organizations such as Intel and the Department of Homeland Security also similarly use pen-test failures as the primary metric of user vulnerability and risk. Thus, organizations, no matter their size or stature, rely on the same metric. Each of them merely counts accidents and nothing more. They not only count accidents but also use the symptom as a diagnostic of user risk. No organization big or small, anywhere in the world, has therefore effectively diagnosed why their users are vulnerable to phishing. Consequently, they all remain vulnerable.

There is a much better way of diagnosing why users are susceptible. In chapter 4, I discussed the cognitive-behavioral reasons people fall for social engineering attacks. Emerging from that research was an important finding, that there is a single mental process that captures whether someone is going to be deceived by social engineering or instead end up detecting it. This mental process is predicated on the same cognitive and habitual user-level factors. Better yet, it can serve as a metric—a diagnostic—for assessing user vulnerability to social engineering. Best of all, it can be easily measured and, as the following vignette of Jenna and Carson will illustrate, it is something we are intimately familiar with.

## THE STORIES OF JENNA AND CARSON

Jenna found a pair of crumpled theater ticket stubs tucked away in her fiancé James's jacket. "That's odd," she thought. James wasn't into theater—in fact, he hated it. Jenna would surely have known in advance of any plans to catch a show. She knew James's friends and they, too, weren't theater buffs, which begged the question "Who did he go to the theater with?" In the ensuing days, Jenna became more watchful. James, who would always call to talk during lunch, wasn't calling as often. He was also dressing up better, and she could swear that he even smelled of perfume when he came back from work, again something that hadn't happened in the years since they first met. Jenna became even more cautious. She began looking for more clues. She checked on what James was doing at lunchtime by calling him, looked over his credit card receipts, and even checked his jacket pockets at the end of each day. She knew something was up. He was on his phone, texting someone more often than he usually did and at odd hours.

Jenna knew the signs. She had seen it before in another relationship when she was a lot younger, and it was happening again. She became vigilant and careful, watching everything James was doing, had done, or was likely to do, looking for clues or "tells" that would reveal what she knew—that she was being cheated on.

My good friend Carson, a yuppie, disliked the way our generation traveled on airplanes and in rental cars. "Maps, messaging, data, all technology was a crutch," he said. His style involved using his good old-fashioned senses and his feet. He liked taking the path less traveled and exploring places on foot and on bicycles, scooters, and public transportation. He window shopped and, some would say, wandered through different streets and past different houses. He got to see communities in action, people at work, at play, living life. He had met people on the main and mean streets of more towns in the US than anyone I know and had gathered a lifetime of wisdom.

The most interesting to me was how he ensured he wasn't in the wrong neighborhood or in a place where he could be in trouble. His general "foot" rule was "Watch for police cars or emergency vehicles. If you see one or more parked, be extra mindful of such areas." Carson believed they were like the "Deer Crossing" signs we see on thruways in the northeastern US. As he explained, "Parked police cars doesn't mean there is trouble there now, but that it has been there and is more than likely! Be cautious, be wary, and watch everything in such areas. Be mindful of your belongings, of eye contact with people, of who is around. Look out for trouble. It's a sure-fire way to avoid it."

There is a common thread that connects Jenna and Carson. It's the mental process of suspicion that was triggered in both of them: in Jenna's case, by the fear of being cheated on, and in Carson's, by the parked police cars and emergency vehicles.

Suspicion is the feeling of unease that is triggered by informational cues in the environment. Cues range from the obvious (police cars and emergency vehicles) to the nuanced (ticket stubs and the smell of perfume). The cues could be visual, as with Carson, to something aural or even olfactory, as with Jenna. They individually or together trigger a connection with something in memory from a time when similar circumstances caused an unpleasant outcome. It is this unease triggered by sights, sounds, and smells in a contextual situation that is suspicion.

Once aroused, it tips the mind to seek information—and a lot of it. This is because suspicion's triggers are rooted in negative events, which humans are programmed to learn from and avoid, so as soon as it is activated, reducing the unease becomes the drive. We pursue this the only way we can: by acquiring information. But because there are experiences to draw from, the information that is acquired is more nuanced and considered. We saw this with both Jenna and Carson, whose triggers were learned in the past. In both, suspicion increased their sensitivity to information. This sensitivity, which enhances information acquisition in volume and quality, enhances the chances that individuals will detect deception.

Suspicion works not just in individuals but also in groups. Think of a jury trial, the bedrock principle of the US criminal justice system, where a group of 12 members representing the voice of the community judge whether a person is guilty. Trial lawyers believe that the longer a jury deliberates, the higher the chances that they will judge in the defendant's favor, finding the person not guilty. Any trial requires that the defense present information and poke holes in the plaintiff's lawyer's presentation to the jury. Remember there is a lot of information that is presented by both sides, usually information that is conflicting and that requires focused interpretation in order to get to the truth of the matter. But nothing that is presented, be it for or against a lawyer's client, matters unless a jury is willing to scrutinize this information.

So how does any lawyer ensure that a group of complete strangers deliberates on information that has no direct bearing to them? A good defense lawyer arouses suspicion in the jury because, once aroused, they are more likely to assess the information.

Suspicion is a human process that all of us have experienced as individuals and in groups small and large. Its influence isn't limited to locating a dangerous street, finding a cheater, spotting a liar, or catching a criminal but permeates all the spaces we inhabit. This includes cyberspace, where it plays a pivotal role in detecting phishing deception. Yet, if you look for published research on suspicion, you will be hard pressed to find much. What you will find instead will be within the realm of mental disorders rather than deception detection. Much of what you will read in security and deception research will be on the role of trust, and there are a number of reasons for it.

## THE ROLE OF TRUST

"Do you think the president is being honest with the American public?" "Is CNN or Fox News a more reliable source for political news?" "Do you think technology companies like Facebook lack integrity?" If you have ever taken part in a public opinion poll, you have likely encountered one of these types of questions. They measure honesty, reliability, believability, and integrity—all different facets of trust. But seldom have you been asked whether you were suspicious of the president, the media, or various organizations. Why, given the pivotal role of suspicion, isn't it asked about or studied more?

One answer can be found in the wordings of the poll questions. Each measures trust because it has various facets or dimensions. In addition to the types of trust just mentioned, there is also credibility, authenticity, genuineness, trustworthiness, sincerity, and dependability. Many of these are influenced by sociological factors (family dynamics, demographics, religion, and culture), making asking about trust more interesting from a public policy point of view. In contrast, suspicion is a singular, fine-grained process. You either have it to a degree or you don't, and its influences—as this chapter later explains—are within the individual, not across sociological factors.

Another reason that suspicion isn't asked about is that trust is more prevalent. Without trust in others, we wouldn't leave our homes, walk our neighborhoods, or drive on roads. We wouldn't go to a restaurant and order food prepared by strangers, expect it to be safe to eat, and pay for it using cash or credit cards issued by some entity we know by name only. The restaurant and its employees wouldn't know whether our money is of actual value and thus whether their work is worth their while. Trust is the glue that makes modern society possible. To ensure trust, we have evolved laws, policies, monetary systems, and institutions such as the police and courts, because of which we are programmed to trust more than distrust. It is a default state. This makes asking the public questions about trust and finding answers about some facet of it easier. In contrast, suspicion is triggered within specific contexts, making studying it outside its context more difficult.

A third reason is that suspicion has a negative connotation. It is considered impolite to ask people whether they are suspicious of a person, political party, or organization. Being suspicious all the time is even viewed as a sign of underlying mental disease. Because of this, suspicion is usually studied within the context of psychological disorders rather than from the deception detection standpoint.

Finally, trust is more stable. Trusting others is part of our core values. Every major religion, as well as many cultures, preaches the value of trust and forgiveness. Because of this, once formed, trust also endures. People are more apt to forgive infractions and maintain trust because of fidelity, loyalty, and faith. The enduring nature of trust makes it easier for us to measure it using surveys and polls that require respondents to recall their thoughts after the fact. This is something that suspicion, which is contextually bound and sensitive to information triggers and individual impulses, lacks. This makes studying suspicion tedious. It requires experiments that carefully and completely simulate the context. It also requires controls to identify the role played by suspicion and what triggered it. Compared to suspicion, trust can be studied more easily using cross-sectional surveys, which is another reason much of the research on phishing has ignored suspicion and focused on trust.

But while trust provides a convenient measure for publishing research findings, you can't gain much actionable knowledge from it. Knowing that someone or some organization isn't authentic or lacks credibility doesn't tell us what led to these conclusions.

Take the case of law enforcement in the US. Going back to at least the early years of the twenty-first century, research on policing has focused on the lack of public trust toward law enforcement.[4] Research has concluded there is a need for increasing trust and has advocated all manner of interventions. Phrases such as "Integrity, Courtesy, Professionalism"—all different dimensions of trust—appear on badges and police cars, and trust building has been the focus of the public discourse. Yet, mistrust of police continues to grow.

This is because it is unclear which dimension of trust is at the root of the problem. Do people mistrust police because they are concerned about their professionalism, because the police aren't courteous, or because the police lack integrity? Are there other elements of trust besides these that matter? Without an overarching framework that incorporates all the aspects of trust, it's unclear what they are. It also makes resolving trust issues difficult, because there are far too many subdimensions that remain unaccounted for and unaddressed. Finally, and more importantly, because of its multifaceted nature, there is no single metric to measure trust. Therefore, there is no way even to contrast the success of different trust-building interventions to know what's lacking. Because of this, the debate on police mistrust has gone on for close to two decades.

Solving a problem requires a measure with a single framework that explains how and why it was triggered. This helps interpret what occurred and how to remediate it. It also helps compare different interventions and calibrate their relative merits.

This is where suspicion once again shines. Trust or mistrust, regardless of their dimensions, triggers suspicion, and there is a single framework that comprehensively explains how suspicion is triggered and what triggers it within different contexts. This framework, beginning with how it was developed within the phishing deception context, is discussed next.

## STUDYING THE ROLE OF SUSPICION IN PHISHING DECEPTION AND DETECTION

How do users differentiate between the authentic and the fraudulent? The differentiation between real and fraudulent occurs in users' minds, as in their cognitions and behaviors (clicking, opening, etc.). Some aid the attacker, whereas others frustrate them. Understanding this process and identifying the user-level factors that aid deception and detection became the singular goals of my decade-long research program on phishing.

As I explained in chapter 4, online cues are mostly visual. They range from the textual (typed words and phrases), to the graphical (everything from GUIs to logos), to the occasional notification sounds and alarms. A variety of such cues are present on websites, in emails, and in applications. They exist on authentic websites and communications and in social engineering emails, messages, and requests. My research simulated different social engineering attacks, each crafted with different cues, and tested them using controlled experiments on different users.

This was a tedious endeavor. It required capturing the within-user cognitive-behavioral processes through which different cues (visual graphics, notifications, notification sounds, interaction icons, buttons, voice) and communications received on different devices (tablets, smartphones, personal computers, laptops), through different attack gateways (email, texting, social media, USB drop-offs, phone calls), lead to deception or its detection. This couldn't be accomplished in one study. It required multiple research studies, each simulating actual attacks on different groups of targets who had opted in and were approved (by human subjects review boards) for testing.

Each study led to the next. Findings served as a launch point and scaffold for refining the next attack and the next study. For instance, in some of my social engineering attacks, college students and adult employees in different organizations were sent phishing attacks that had only textual information in the emails (much like a pretexting Nigerian-type attack) or alternatively had only graphical cues. This helped identify whether cues even mattered or whether text alone in emails sufficed. Another set of studies spoofed or created fake websites and form emails of well-known brands and, keeping the content the same, altered various visual elements such as the color, logo, and graphics. Incorporated into the visuals were the cues from the earlier study, as well as measures of how and what users were thinking and doing. In this way, each study helped pinpoint which visual elements triggered the most reactions, as well as how they did so.

Once the most important visual cues were identified, the next research studies incorporated different textual messages along with cues, with the goal of assessing whether there were certain cue and textual combinations that worked better at deceiving users. For instance, one attack announced a lottery, while another provided a deadline and warning to comply with a request.[5] Both simulated attacks had the same logo and colors of a well-known organization. Other such simulations examined social media–based attacks, again iteratively changing different textual, graphic, and message elements. There were also simulations of faked versus real login pages of well-known shopping websites, unfamiliar websites, social media sites, email services, messaging services, cloud sharing portals, Wi-Fi gateways, VPN services, and other major and minor attack vectors or ruses that social engineers use.

The simulations had to be carefully designed. All used controlled experimental conditions with an equal number of opposing signal and control groups. For instance, if there was a spoofed social media login page that had a typographical error (say Facebook was called "Fa ebook") and there was another with a color that differed from the blue used in the official Facebook login page, then there was also a login page with the typographical error that used the different color and another with the correct, authentic login page that served as a control. (Such balanced designs—called factorial designs—allow examination of the interactions between multiple cue combinations using a single experiment.) None collected personal data, such as the participants' logins or passwords, and care was taken to ensure everyone consented to participating. Every participant was also debriefed at the end and provided with knowledge and training on how social engineering worked and what their data suggested. None were treated as guinea pigs—they were users, whose information was going to help stop social engineering.

Users' reactions to each attack were tracked using behavioral data, such as whether they had opened their email; how long they had looked at a webpage; whether they had entered their logins on spoofed pages; the number of people they had contacted, reported, or followed up with; and whether they responded to messages. Users' reactions were also assessed using follow-up surveys that employed a range of cognitive and behavioral measures that captured what the users were thinking and the extent to which they were being honest in reporting their behaviors. This provided two layers of data: what they did and how they thought. Some of the data overlapped (e.g., tracking email clickthroughs and users' responses to the survey question asking whether they did), which helped in understanding whether the user was being honest in their responses. This helped ensure that the conclusions reached were valid.

This took years, and it wasn't without many misses. Some studies led to dead ends. For instance, a user's personality factors, gender, and geographical location didn't matter. The persuasiveness of text in emails, the spelling in them, and even factors such as curiosity mattered little, but they were important signposts. They directed the research to its ultimate destination—unraveling the mechanics of deception or its detection within users. The iterative research process ensured that the mechanism identified through this process exemplified what mattered most in the attack and

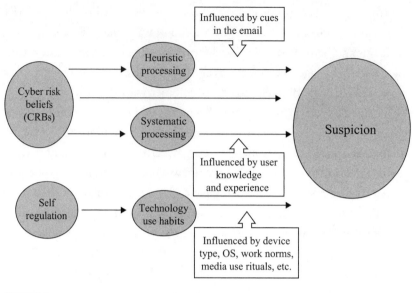

**Figure 5.1**
Suspicion, Cognition, Automaticity Model (SCAM).
*Source*: Vishwanath et al. (2016).

within the user. It helped pinpoint why attacks succeed, how users react to them, and the significant cognitive-behavioral predictors of their reaction.

All these interrelationships and their influences could be captured by a single research framework that comprehensively explained how users were deceived by a variety of phishing attacks. I called it the Suspicion, Cognition, Automaticity Model (SCAM), and it is illustrated in figure 5.1.[6]

### WHAT THE SCAM FRAMEWORK SHOWS

Overall, the one user-level factor that consistently emerged in the SCAM framework was that users who were suspicious were significantly less likely to fall victim to social engineering attacks. Suspicion was triggered by supporting thoughts, ways of thinking, and actions. They ended up protecting users from the deception.

What were those supporting thoughts and actions? They were the very factors we discussed in chapter 4: cyber risk beliefs (CRBs), the two modes of cognitive processing, and users' technology habits and their personality-based precursors. To recap, CRBs are users' beliefs about the inherent risk of their online actions. The two modes of processing they motivate are heuristic processing, which involves the use of cognitive shortcuts triggered by cues, and systematic processing, which involves elaborated thinking about the elements in a message. Habitual responses are the automatic, nonconscious reactions that are triggered by rituals and patterns of media usage. These have a basis in the self-regulation or self-control users can exercise when online.

How these influence user deception and detection is depicted in the model, which is read from left to right, with influence flowing from CRBs and self-regulation via their impact on cognitive processing and habits. The cognitive and habitual factors most directly trigger suspicion. They are therefore the immediate or proximate factors in the influence chain. CRBs and self-regulation stem from longer-term forces such as prior experiences with technology and with life in general that lead to the development of personality and beliefs about technology. They are the ultimate factors in the model.

The proximate factors are also contextual, in that situational variables influence them. Cues that are presented in the attack (as in their *availability*) and their relevance to the user (as in their *accessibility* in memory) directly trigger heuristics in the user's mind. (We discussed why they matter in chapter 4's discussion on heuristic processing.) Knowledge and prior experience about social engineering affect the quality of systematic processing. Using smartphones and concomitant factors such as the operating system used, the work norms regarding how files are transferred within an organization, or media use rituals (such as combining different email accounts into one mail app or not wanting to see unread email counts) influence the degree to which habits manifest.

All these processes occur in mere seconds in response to a conscious reaction or nonconsciously with habits. The factors within the model also recursively influence each other. This means that although the model is shown as flowing in one direction and happening just once, the reality is otherwise. We think, see evidence, and then think again based on what we have seen. These occur in loops. In response to suspicion, users might engage in a cognitive effort to discover deception, which could increase suspicion and might in turn engender greater processing and even more suspicion. Such feedback loops are implicit in human thought processes. The model displays the most salient paths at the culmination of this process.

Ultimately, what matters is whether these paths trigger suspicion. The arousal of suspicion protects users because it changes how they relate to an attack. Across the research studies, in response to an attack, users who had high levels of suspicion possessed stronger risk beliefs, they knew to focus on the elements of the attack that mattered, and they had acquired supportive technology use habits. These protected them but not just from that attack. Because strong risk beliefs and good technology habits develop over time, their protection acted as a barrier against similar attacks. They served as a security blanket, making users resilient.

The internal subprocesses influencing suspicion worked in a couple of different ways. With most users, the influence of CRBs was indirect, usually by motivating cognitive processing. Take, for instance, an encounter with a spoofed shopping portal that asks for a victim's social security number. Users who believed that providing this information would put them at risk usually followed up with a detailed examination of the request's authenticity. They usually changed their cognitive processing strategy, engaging in systematic processing, and erred on the side of rejecting anything that the email attempted to garner from them. With some users, strong CRBs directly triggered suspicion. This occurred when because of past experience users already believed something was suspect. For example, when warned about the IRS phishing phone-call scam , many users, already aware of such attacks, immediately became suspicious and hung up. Thus, in the presence of domain knowledge or prior experience, the activation of CRBs and the suspicion they triggered led users to detect a variety of phishing attacks.

Conversely, users who didn't become suspicious fell prey to attacks. This occurred more often when users heavily relied on cues for decision-making. This was usually because the associated heuristics they triggered were suboptimal owing to all the issues we already discussed in chapter 4: the preponderance of cues online, similar cues triggering different heuristics, and different cues triggering the same heuristic. Social engineering attacks where cues were injected strategically and took advantage of one of the preceding issues were especially likely to lead to a relaxation of suspicion and succeed in deceiving users.

Suspicion was also not triggered when users reacted habitually. Habits, again as discussed in chapter 4, are acquired online behavioral patterns and rituals that are enacted with little conscious forethought. They are predicated on the heavy use of smartphones and include patterns that are defined by device use, such as combining different emails on a device, and by personal rituals, such as not wanting to see unread notifications or checking social media at work. Such patterns foster reactivity to social engineering attacks and vitiate the user's ability to pay attention to the message elements that could trigger suspicion.

Thus, overall, suspicion is a key symptom. Because it is determined by attack-level factors (cues, persuasive text), user technology (device type), and user-level factors (CRBs, cognitive processing strategy, habitual patterns), suspicion singularly accounts for the mechanism of user deception via phishing. Whenever suspicion isn't sufficiently aroused, users fall prey to social engineering. When it is triggered, users tend to detect deception. In other words, measuring suspicion reveals why users are vulnerable to social engineering.

## SUSPICION: THE SINGLE MEASURE OF DECEPTION AND DETECTION

Measuring the extent of suspicion triggered in any user provides a simple, direct mechanism for diagnosing whether a user is likely to fall victim to social engineering, for all the following reasons.

First, suspicion does not require multiple measurements. In the stories of Jenna and Carson presented at the beginning of this chapter, suspicion was aroused in both, but compare this with trust, the factor that most researchers prefer to measure when examining deception. In Jenna's case, her suspicion was triggered by a need for honesty and integrity in romantic relationships, while Carson's was owing to a different facet of trust, his trust in societal institutions, such as the police in maintaining public safety. These are three different typologies of trust, and there may be more underlying them. There might be James's trust in ethnic groups or Jenna's trust in male partners, which the measurement of the more apparent three facets of trust does not capture. However, even if we were to measure all five facets of trust, the issue would become one of comparability. The same facets of trust weren't experienced by Jenna and Carson—which means the measures conflate the different types of trust. In contrast, suspicion is a singular process. We can capture it with a single simple measure and understand both Jenna and Carson's mental processes.

Second, suspicion is triggered across a variety of contexts. It is therefore relevant as a measure across different attack contexts. This is important because social engineering attacks encompass different media. They primarily use email and messaging platforms but also use phone calls, USB sticks, Wi-Fi spoofing, and other means to gain access to users or their credentials.

Third, the antecedents of suspicion are not in technology but rather in users. Suspicion's ultimate influences are localized in the users' minds. They are in users' habits, their internal regulatory processes, cognitive processing approaches, and their cyber risk beliefs. When we measure suspicion, we account for all these. Technology's impact on deception occurs through the localized influencers, which means it's accounted for as well. It also doesn't matter what technology is available today in a given organization or what it evolves into tomorrow—the measurement of suspicion doesn't have to change to accommodate it.

Fourth, suspicion is a very sensitive mental process. As we saw with Jenna and Carson, a small amount of suspicion is all that is necessary to set it into motion. This makes the measure ideal for capturing even slight changes or slips in the contextual environment that might reveal deception. This is necessary because many social engineering attacks, especially the more sophisticated ones, are well designed. Measuring suspicion allows us to capture the impact of even the most nuanced elements of these attacks that cause deception.

Fifth, once triggered, suspicion leads to a disproportionate increase in information acquisition and related behavior. Measuring it therefore allows us to assess not only how vulnerable any person is to a social engineering attack but also their likely reactions to it. We could identify the types of content users are likely to seek, where they might look for it, and how to package it to meet their information appetite. We can then help make such solutions available where we know the user would look for them.

Sixth, suspicion reoccurs in different analogous situations. As with Jenna, once a romantic partner has experienced infidelity, they are all the more cautious in their next relationship, even discovering signs of cheating quicker. The same goes for people who have been unfortunate enough to experience crime on a city street. They become, as the saying goes, "once bitten, twice shy." Suspicion is therefore a rather good predictor of how someone is likely to think and act in the future under similar circumstances.

Additionally, because the triggers for suspicion are learned through prior unpleasant experiences, to avoid repeating them, its cues get firmly imprinted and readily accessible in the user's memory. This makes it possible for people to recall the specific cues they relied on with great accuracy—something that both Carson and Jenna did. Thus, the singular measure of suspicion can be used not only to determine who is likely to fall for a social engineering attack but also to identify the cues they missed that should have triggered suspicion. This knowledge can be used to develop novel interventions that fix such flaws at their source, in users' cognition or behavior.

Finally, because suspicion can be reliably recalled, it can be measured on a self-report basis. Because of its singular manifestation, we can measure it using a standard *not at all suspicious* to *very suspicious* survey question. This makes it possible to capture the degree to which suspicion was evoked in a pointed and direct manner. It's in sharp contrast to the existing approach of counting pen-test failures that provide dichotomous pass-fail data.

Thus, measuring suspicion explains the how and the why of user vulnerability. It tells us how users related to the cues in the attack and which cues and internal processes led to deception or its detection. This explains why some fell for the attack and why some didn't. It also helps us predict who is likely to fall for similar attacks. By explaining and predicting, suspicion allows us to identify the weaker links in need of protection. By telling us about the how and who, it allows us to develop solutions that work for them and have them available where users seek them. In this way, measuring suspicion allows us to diagnose users' vulnerability to phishing. How we do this is explained in chapter 6.

# 6  PERFORMING AN ACCURATE DIAGNOSIS

The stethoscope, invented over two hundred years ago, is a simple measurement tool that can be used for diagnosing a whole range of minor and major ailments, from respiratory infections to abnormalities in blood pressure. Used correctly, this $15 device can save lives, but it's easy to use it incorrectly.

When measuring blood pressure, a pressure cuff is placed on the upper arm of the patient by a physician, who simultaneously places a stethoscope on the patient's forearm and listens to the sound of their pulsing blood as the pressure from the cuff is released. If the cuff were too large or small, placed too high or too low on the arm, or if the patient were lying down instead of sitting, the overall pressure the cuff would exert would be inaccurate. The reading on the stethoscope would be off. This could mean the difference between misdiagnosing a life-threatening ailment or providing timely care.[1]

Medical practitioners spend years in school learning the correct procedures for using measurement devices such as stethoscopes. It begins with an understanding of human anatomy and the procedures for simulating different symptoms. In the same way, IT managers need to understand why users are susceptible to deception based on social engineering (which we covered in chapters 4 and 5) and learn the procedures for assessing the symptoms in their users. Only then can they perform an accurate diagnosis.

We already discussed in chapters 4 and 5 the anatomy of attacks and what makes users vulnerable to them. Now we examine the rest of the procedures necessary for accurate diagnosis. To begin, we need to trigger the symptoms leading to social engineering–based deception or its detection in users. This has to be done in a manner where we can track what our patient—the user—is thinking and doing. The easiest, most widely applied approach is to conduct a social engineering penetration test (pen test).

Pen tests can vary considerably. They can simulate spear phishing (as in emails carrying hyperlinks or attachments) and pretexting emails (as in some ruse that leads to a back-and-forth interaction), social media impersonations, USB drop-offs, Wi-Fi spoofing, or any other social engineering scheme. Attacks can also range in modality (e.g., an email, a text message, or a phone call); in deceptive approach (e.g., they could announce a reward or could involve a spoof of a website or Wi-Fi login process); in the payload they carry (e.g., have a hyperlink or an attachment or simply a persuasive text message); in the amount of personalization (e.g., they could name the user or come from either the company's IT department or a well-known brand); and in deceptive intent (e.g., they could trigger macros in programs, open a spoofed page that collects credentials, or contain a tracking cookie that tracks users). Regardless of the approach, the goal of the pen test should be clear to those designing it. It should be to trigger the symptoms in users that accurately reflect how they would react to a social engineering attack.

This isn't a trivial issue. IT staff who design a pen test often don't focus on its quality from a diagnostic point of view. Instead, most create pen tests in an ad hoc manner. This is not just within organizations that craft their own tests but also in corporations that use established cybersecurity training packages or employ consultants.

Pen tests are developed ad hoc because there is no real benchmark for what an "appropriate" phishing test entails or what an ideal phishing test must contain. There is no framework, let alone any guidance, on how best to develop a test or analyze its appropriateness. For instance, is an email with an attachment as payload the same as an email with a hyperlink?

Is an authentic email from a hijacked account—a vector that most pen tests avoid simulating—not relevant? Is an email that spoofs a well-known online retailer comparable to a spoof of the cloud service that the organization utilizes? Is an IRS email sent after the tax season comparable to one sent during tax season? If so, are they comparable enough for us to assess user resilience across these attacks? The answers to many such questions just don't exist. Notwithstanding this, training data that uses all these types of attacks is used to assess user readiness. Worse yet, the data is used both within and across organizations to inform policy, even by cybersecurity insurers for their actuarial estimations.

Instead of addressing the fundamental issue of pen-test appropriateness, cybersecurity training has focused on achieving an outcome—and in this case the outcome is to see how many users fall for a pen test. Many make it a game, with the goal being to trick the user. For this, they instill fear, which as we witnessed with the National Accounting Agency's users in chapter 4, ends up backfiring, as users figure out ways to avoid the test.

With no real framework to guide their pen-test design, IT staff often look for exemplars online or use databases of prior attacks. Many have to reuse tests because pen tests are often conducted throughout the year, more than once each month in some organizations. But tests that replicate prior attacks are usually known to users, so it's possible that users are already aware of them, having seen them in the news or online. Even if this isn't the case, the success rates of most social engineering attacks are unknown because they are not tracked objectively. There is no organization that tracks how much success an attack has achieved, let alone what in it made it successful. With nothing to connect an attack's success to its features, there is no way to know why when it was reused in a pen test it worked the way it did.

Even those in IT departments who reach out to security vendors for guidance on test creation fare no better. They are given pen-test exemplars and arbitrary guidelines. In one anecdotal case, an IT manager who sought guidance from a cybersecurity vendor from a well-known security training firm was told that an ideal test was one that achieved a 10 percent failure rate—because that is what they had seen across organizations. Besides the fact that there is no scientific basis for this requirement, this example underscores the problem with pen testing. Rather than focusing on the quality of the test, the focus is always on how well it can trick a certain percentage of users. It is therefore no wonder that in organizations worldwide, IT staff keep designing pen tests regardless of their quality or appropriateness.

Because of this, today there are all manner of different pen tests being deployed. Some tests are so amateurish that everyone who receives them detects the deception. Others are so well crafted that they do not differ from legitimate emails, so no users can detect them. Some are entirely irrelevant to users (e.g., offering fake scuba diving lessons to users located in an area with no water bodies nearby); some others use a mishmash of company names, logos, and colors; still others are text-only with varying appeals, typographical errors, and deceptive content that range from the nuanced to the obvious. The varying and unknown quality of pen tests renders almost all the data on user vulnerability that has been gathered until now by organizations all over the world wholly invalid. As a result, we cannot trust any of this data.

The root of this problem is the lack of a scientific standard for what constitutes a phishing pen test. This influences any diagnosis or conclusions that follow from it. Extending the earlier health care analogy, it is similar to there being no guidance on where, when, or how to place the blood pressure cuff or stethoscope on patients. Thus, no matter the quality of the stethoscope or cuff we utilize, the blood pressure reading itself cannot be relied on for diagnosis.

In chapter 5, we discussed how suspicion, derived from SCAM research, serves as an appropriate measure for user vulnerability to social engineering. Just as a good stethoscope alone doesn't guarantee an accurate diagnosis, measuring suspicion alone doesn't suffice. We need to measure it in response to a scientifically valid baseline phishing pen test. The next section lays out how you can design such a test.

## A FRAMEWORK FOR DEVELOPING VALID PHISHING PEN TESTS

Recognizing the absence of a guiding framework, part of my research focused on developing it. Given the scope of this book, the focus here is on an email-based pen test, but the framework presented here can be extended to address other forms of social engineering as well.

The framework for assessing the quality of phishing pen tests, called the Vishwas Triad or simply V-Triad, was developed in 2016. I presented it at Black Hat USA in 2016 and 2017, and a summary of the development process is presented here. (For anyone interested in learning more about it, a video of my presentation, along with a white paper detailing the research that developed the framework, is available online.[2])

To develop the framework, my research team gathered five hundred or more real phishing tests from different organizations. Organizations provided all their exemplars, as in the pen-test emails they deployed, as well as the resulting user failure rates. The data covered a range of attack types, persuasive approaches, and sophistication. The failure data covered user clickthroughs, detailing who, how, when, how quickly, and from where they clicked on each email and failed the test. The number of users subjected to pen tests ranged from a hundred to close to a thousand, so the cumulative subject pool of participants was large. Many of the emails were similar to real-world attacks or were facsimiles of actual attacks (because of the practice that I alluded to earlier, where IT staff search online for different pen-test ideas). To the extent the attacks replicated real-world phishing attacks, we could consider them reflections of how social engineers think. The research process for developing the framework inverted this thinking. While social engineers launch attacks and collect data based on their success, I used the data on how successful the attacks were and matched it to the features that made them successful.

To understand the relationship between each attack's features and its success, the research used a methodology called content analysis. This is a quantitative approach to analyzing information that involves categorizing email features into nonoverlapping, independent categories through a process called coding. Coding in content analysis uses human coders who independently inspect each email, count the presence and frequency of various features in them, and place them in different categories. The process quantifies the various structural features, their presence, frequency, placement, timing, and quality. Multiple coders inspect the same emails, which allows for a comparative assessment of their accuracy. Moreover, none are aware of the success or failure of the pen test. Their coding is focused solely on the cues in each email. After all the emails are coded, the data from the coding process is arrayed with the user failure data and the relationship between the feature sets and the clickthroughs are statistically assessed.

The analytic process culminated in an important revelation: almost everything in the phishing emails that mattered were those that either signaled trust or suppressed suspicion. Attacks that did this successfully netted more clickthroughs in the shortest amount of time.

As discussed in chapter 5, trust, unlike suspicion, has many dimensions or facets (credibility, authenticity, believability, personality, and so on). Successful phishing emails incorporated one or more of these because there were more trust signals available. For instance, high-clickthrough attacks either had a well-known brand's name or logo, suggesting the email was from a trustworthy and credible source; incorporated personalized messages to suggest the email's persuasive content was believable; or included phrases such as "Sent from my iPhone" to show an email was from an actual person rather than an automated email. Often these were combined and had a cumulative effect in reducing suspicion. High-clickthrough attacks also incorporated familiar form fields and persuasive refrains, better timed their attack, and preyed on users' routines and beliefs. These worked in tandem to suppress suspicion.

To understand how all these worked together, the analysis compared the individual phishing email features to their respective clickthroughs and, working upward, grouped the features into broad categories. This led to three nonoverlapping categories within which all the features fit. The categories served as higher-order predictors of the major components of phishing emails most likely to cause deception. They were organized into a visual framework that showed how the three predictors worked together to instill trust or reduce suspicion. The framework's name was chosen to reflect the social engineer's intent. It borrowed from the Sanskrit word for trust and was called the Vishwas Triad or simply the V-Triad. The model, along with the higher-order predictors, is presented in figure 6.1.

Each vertex in the V-Triad corresponds to a major deceptive feature in phishing emails. The two-headed (i.e., bidirectional) arrows signify how the essential facets of spear phishing emails interact with each other and enhance deception. One vertex signifies the email elements that indicate trust: *credibility signals*. Another captures the aspects of the email that are modifiable; let's call them *customizability signals*. The third vertex involves the applicability of the email to the user, or its *compatibility*.

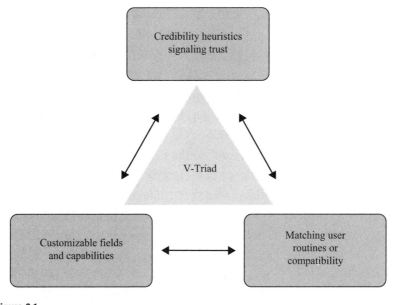

**Figure 6.1**
The Vishwas Triad.
*Source*: Vishwanath (2017).

Credibility signals can be icons, cues, or even textual elements. They can range from the obvious, such as using the name of a known person, to the subtle. For instance, the phrase "Re:" in the subject line might subtly indicate that an email is part of an ongoing email exchange. Because of the many elements that exist in online communication—cues, texts, graphics, drop-down menu elements—there are more opportunities to signal credibility and for them to be cumulated.

Emails provide many more avenues. Credibility signals can be introduced into the sender's name, their email address, subject line, salutation, signature, body text, or attachment fields. As discussed in chapter 4, they could be impactfully communicated by a PDF icon in an attachment, because many users mistakenly believe such document types are hackerproof, or by the inclusion of an SSL icon, because many users mistakenly believe it is a sign of a website's authenticity. Credibility signals work because they are consistent with the available heuristics in the user's mind, or they tap directly into a cyber risk belief. Such signals are thus more successful in deceiving users when the users have experienced them during email exchanges.

Customizability signals are facets of online communication that users have modified or have seen modified. Unlike credibility signals, which feed on consistency, customizability creates comfort with changes in online communication. It relaxes suspicion. For instance, all online users have changed subject lines in emails and expect them to change. Similarly, URLs change when we navigate beyond the home page of a website, and many such URLs also contain unusual characters. We also discussed in chapter 4 how in many cloud services, while the email appears to come from the sender and has their name, it doesn't come from their in-box. Instead, it comes from a different in-box, one that changes with the provider. Such variances are part of most users' online browsing experiences and extend even to the sources of emails and the addresses from which emails are sent.

In other instances, because of how email clients such as Gmail are programmed, emails are organized by conversation, so the same subject line keeps getting inserted by default, even if the content of the conversation may have changed. Users are constantly accommodating such changes by overlooking the subject line or ignoring its mismatch. They are also used to disregarding lessons about spear phishing because legitimate email exchanges frequently violate even the most basic tenets of email security. A telling case is a 2013 email from Target's CEO, Gregg Steinhafel, informing some 50 million customers about a breach of their personal data. The email contained all the deceptive signs of spear phishing that users are cautioned about. It was sent from a "TargetNews@target.bfi0.com" in-box with which no customer of Target was familiar. It even contained a hyperlink accompanying an offer of free credit monitoring and identity protection along with a deadline for enrollment.

In addition to accommodating such variances, users are also accustomed to entering information that appears on different devices in form fields that also constantly change. From a spear phishing perspective, the most pertinent are credentials—logins, passwords, and two-factor authentication (2FA) codes. On many new online services, users are allowed to log in with a single sign-on (such as using Facebook or their Gmail account credentials). This requires accommodating multiple authentication windows. For instance, when signing into a new service, the user must enter the credentials from the single sign-on service provider on that website, then authenticate their use with the sign-on provider, then revert back to the new service's website, accept its terms of contract, and set it up. Each authentication prompt varies based on the website and the sign-on provider.

The same is the case with 2FA codes. Some services, such as Yahoo Mail, use a specific code color and layout. Others, such as Apple, have a device-based authentication, where the prompts show up on different Apple devices that also include the location information of the device asking for the code. Many banks and online services present the prompts via a text message or a phone call.

None of these is static in design or in terms of the code. Each evolves based on the app, the service, the device the user is using, and their respective locations. Because of the many variations, online users often discount something that might look different or inconsistent from what they usually experience. They chalk it up as another customized variation—and then comply. Such facets of emails are captured in the second vertex of the V-Triad.

The third vertex of the triad is the email's compatibility, or relevance to the user. This encompasses the elements of the email, from its content to its structure, topic, and timing, and how they fit with the norms, work patterns, and routines that the user has been socialized into through technology use. At the simplest level, this would be whether the user who receives a phishing email purporting to be about, say, their Instagram login even has such an account. At its more advanced level, compatibility comes from the fit between what the user has seen or experienced and what the user expects within similar communication contexts.

For instance, someone working in the accounting department who often receives requests for user information files via email from the CEO might not find it odd to comply with a phishing email asking for the salary data of all users in the organization, even though such a request might never have been made in the past. This is because the norms of the task have routinized such requests and the accounting person has blindly complied.

Likewise, if users are routinized into entering their email credentials to access the Wi-Fi at the Starbucks they visit, then such requests, even when they appear in a different location, might be complied with. As an example of how this works, in 2019 the Wi-Fi access prompt of a hotel I checked into in Washington, D.C., asked for a Gmail or Yahoo email address— something that had no relevance to the task. The password field required that the hotel room number be entered. I inquired about this at the front desk and was told that no one had ever complained about it. Every guest had simply complied. It's more than likely that most had even entered their email password until they realized that wasn't what was expected.

Of course, not having an Instagram account, not working in accounting, or never having used Starbucks's Wi-Fi would diminish these effects, but, as my example shows, compatibility tends to flow across our experiences online. While we may not have encountered such requests at work, routinization can occur from encounters outside work. After all, many of the communication devices, services, and apps are used at home as well, so the expectations for what is routine can be created anywhere. Compatibility signals cause deception whenever such expectations are met and when violated they don't.

## USING THE V-TRIAD TO DEVELOP AN IDEAL PHISHING PEN TEST

The V-Triad presents the top-level view of the structural features of phishing emails that lead to deception. Multiple subcomponents or feature sets within phishing emails fall within each vertex. Sample features are laid out in the sections that follow, "Credibility Signals in Phishing Emails" and "Customizability Signals in Phishing Emails."

In most effective phishing attacks, the components from within each vertex work together. The interaction between them increases their overall effects. As an example, let's take a spear phishing email that purports to contain a series of PDF files that are accessible via a Google Drive hyperlink. The hacker might inject credibility cues in the subject line of the email with just the word "Re:" that would cue the user to think this was a continuing communication. The body of the email may contain the colors and layout of

**Credibility Signals in Phishing Emails**

- well-known brand name (Apple, Amazon, Instagram);
- name of friend, colleague, family member;
- photograph of known or familiar individual;
- colors, fonts, icons, and text combinations that mimic known or familiar brands;
- graphical cues indicating type of attachment (Word, PDF, etc.);
- domain spoofing (i.e., copying known websites or URLs);
- using cueing words or phrases (e.g., "Sent from iPhone," "warning," "password," "deadline");
- count of shared connections on social media;
- obvious spelling errors and typos; presence or absence of receiver's name in request;
- security indicators (e.g., SSL icon);
- words such as "Re:" or "Fwd" in email subject line or body; out-of-office notifications; elements of the graphical user interface (e.g., login, reply, accept, send).

**Customizability Signals in Phishing Emails**

- subject line, form fields (e.g., search bar, 2FA input form, login input windows);
- URLs (including from cloud services);
- emails from within cloud services;
- email addresses of different senders;
- sender's name;
- social media updates, email subject lines, prompts (e.g., for accepting cookies, terms of contracts);
- layouts and colors of certain login pages, multifactor prompts and layouts for some services;
- login notifications (informing where and when someone logged into a service), varying single sign-on options on websites;
- changing styles of prompts requesting access to files, folders, settings (e.g., request to enable macros in Word); graphical elements such as memes or emojis.

Google Drive and indications that the file being attached is a PDF. Such an email would be easy to create. It doesn't need a subject line or even a message in the email's body. It can be sent from any email address because Google Drive doesn't allow for any personalization, and hyperlinks are usually shared from an internal in-box of Google's choosing. Having encountered similar emails from the same or similar file-sharing programs, users are likely accustomed to this, which induces a high degree of trust in such emails.

**Compatibility in Phishing Emails**

- mimics a work-related process (e.g., printer sharing, cloud service, internal emails);
- mimics a public occasion, holiday, or event (e.g., Christmas shopping or tax season);
- timed with when users have breaks (e.g., lunch time), when they are more likely to check work email (e.g., Tuesday and Friday mornings), when they are more likely to check email on mobile devices (e.g., Thursday evening in the summer, late Friday evening and night);
- replicates life events, interests, and circumstances (e.g., pregnancy, pet ownership, political affiliation);
- mimics a routine (e.g., checking social media in the morning, paying credit card bills at the end of a cycle, lottery purchases, logging onto Wi-Fi in public places);
- has been the subject of targeted training (e.g., password change emails from IT department, phishing pen-test emails);
- mimics an update to a patch, definition file, or software.

Using the V-Triad, you can easily locate each of these specific elements and the source of their impact on the user. This is what makes it so powerful. Using it, IT staff can not only pinpoint the specific deceptive elements of phishing emails that make them virulent but also design effective pen tests that incorporate all the known elements that lead to phishing-based deception.

The V-triad derives its explanatory power from the Suspicion, Cognition, Automaticity Model (SCAM). As you'll recall, in chapter 5 we discussed how the SCAM describes the proximate and ultimate factors that determine phishing vulnerability. Ultimate factors include cyber risk beliefs and personality-level factors (self-regulation) that via the intervening proximate factors of cognitive processing (heuristic and systematic processing) and habits trigger suspicion.

The impact of each feature set within each vertex of the V-Triad can be understood based on how they influence one or more of the SCAM's factors. Various credibility cues in emails, described in the V-Triad, reduce cyber risk perceptions and trigger accessible heuristic shortcuts. Customizability signals, by accommodating variance, reduce risk perceptions; they foster mental shortcuts and reactive behaviors. Compatibility reduces risk perceptions and, if the communication is routine, triggers habitual reactions.

In this way, the V-Triad explains why different elements are injected into phishing emails and why they work well. It explains why phishers use well-known brand names such as Microsoft, Facebook, LinkedIn, PayPal, the IRS, and others. Having a well-established brand allows the use of credibility signals that are familiar to a lot more people. They trigger heuristics that are widely available and easily accessible in users' minds.

The V-Triad also explains the relative success of some attacks over others. For instance, in my own simulations, phishing attacks that spoof a poorly crafted LinkedIn web page net a higher percentage of victims compared to a poorly crafted Facebook website. The reason is that users of LinkedIn experience more changes on their login page than in Facebook. This routinizes them into entering their credentials on a constantly changing login page. Thus, while Facebook may have more users available for targeting, successfully netting the same percentage of victims on Facebook requires a closer replication or spoof of its services.

The V-Triad also explains why certain attack types have limited deceptive value. For instance, attacks that reward users with lottery tickets only work on the subset of users who have purchased lottery tickets. Most people seldom win a lottery or believe they will; their risk beliefs aren't supportive. Likewise, emails invoking curiosity have a short-term influence. Their influence wanes rather quickly as users become more aware of the attack and form heuristics that foster avoidance (e.g., *if* it is too good to be true, *then* it likely is).

This was the fate of the original Nigerian pretexting emails, which at first succeeded by invoking curiosity and then quickly lost their effectiveness. The same is the case with emails that warn users about account closures, threats of IRS liens, and credit card defaults. They may invoke credibility and might even be compatible in their early stages when users aren't aware of such attacks, but as soon as news of the deception becomes widespread, the attacks become incompatible. This is also why phishing attacks keep changing all the time. They do so because no single attack is effective all the time as users learn about it, form new heuristics, and figure out its incompatibility.

The constantly changing social engineering attacks underscore the value of the V-Triad's approach. Its explanation isn't limited by cues, form fields, and persuasive texts that may change as attacks evolve. Instead, because it focuses on the intent behind them and their paths of influence, it helps in understanding what contributes to any phishing attack's deceptiveness. This helps in estimating the potency of different attacks and in designing pen tests that appropriately replicate them.

So, what is an ideal phishing pen test, and how does one design it using the V-Triad? An ideal phishing pen test encompasses elements from each major vertex of the V-Triad. This is necessary to ensure that all the known triggers that lead to deception in most cases are in it. It means having cues that are credible, having signals that at a minimum invoke customization and definitely don't violate it, and ensuring that aspects of the email are compatible with the user.

But what should those specific cues be? The V-Triad doesn't specify what they should be. This is because individual facets within any attack can and—given developments in technology and user awareness—do change along with apps, services, operating systems and other software, practices, processes, norms, and awareness levels. The ideal cues are therefore those that are suitable to a user. There is no generic, one-size-fits-all attack that works for everyone. We know this because hackers have tried to create them. The infamous Nigerian pretexting email is a glaring example of an attack that has tried the same approach—and we know it doesn't work. Thus, an ideal attack is one that works for a specific group of users in a specific organization or sector at a specific point in time. Its suitability is empirically determined, meaning it is ascertained by measuring its impact on a subset of its intended targets. This requires quantifying the pen test's suitability.

The good news is that because all social engineering attacks attempt to foster trust or suppress suspicion, the same measure of whether suspicion was aroused on its targets can also be used to measure a pen test's suitability. This makes it easy to quantify whether an attack is ideal as well as diagnosing why users got deceived—because it can be achieved using the measure. And it's even simpler than that. All it takes to assess them reliably are a few questions. The next section presents how these questions were identified.

## DEVELOPING THE DIAGNOSTIC

*Can you describe on a scale of 0–10, where 0 is no pain and 10 is severe pain, how much pain you feel?*

*On a scale of 0–10, where 0 is very well and 10 is very unwell, how do you feel today? Can you explain why?*

Such simple questions are a pivotal part of any medical practitioner's toolkit. They are the reason medical science is more effective in humans than in animals. Human patients can be asked questions, the answers to which can help identify and remedy their ailments. Physicians spend years in their medical training learning not just to diagnose patients but also to ask the right questions.

We social scientists also rely on questions posed to research subjects who serve as proxies for our patients—the larger human social grouping or population our research is attempting to understand. Our questions are usually administered indirectly, through the use of surveys, which the respondents provide answers to by themselves. We call this self-reporting. But because we aren't physically present when the respondent answers these questions, the chances of misunderstanding them and of bias and errors creeping into the responses are rather high. Because of this, asking the right questions and asking them correctly is even more pivotal in social science research. Most academic scientists spend years in graduate school learning how to create self-report questions.

The self-report questions they learn to craft are of two types. One, for which the respondents' answers are bounded or structured, is called a closed-ended question. These are the questions that are commonly encountered in online polls with various numerical rating scales, such as the 1–5 response scale where 1 = *strongly disagree*, 2 = *somewhat disagree*, 3 = *neither agree nor disagree*, 4 = *somewhat agree*, and 5 = *strongly agree*. The other type of self-report is the open-ended question, where the respondent provides their free-ranging views or thoughts on a topic, unbounded by a response scale.

The roots of both types of self-report questions can be traced to early twentieth-century research on human psychology. Many of the original theories of the human mind were developed using open-ended questions.

Answers to them provided the framework for subsequently asking more-direct questions. This led to structured closed-ended questions and, with the development of better statistical analysis tools and techniques, the 1–5 type of scaled question used today. For instance, Sigmund Freud's work began with unstructured interviews with patients about their childhood and parental experiences, work that took time to develop and required significant insights into human behavior to recognize the common patterns in their responses. This work gave rise to tests like the Freudian Personality Test, a battery of around 50 structured, closed-ended questions that individuals respond to using a *strongly agree* to *strongly disagree* rating scale.

Over the next century came even more self-report tests, from those measuring disorders to those measuring intelligence. Most started off as open-ended questions and were in time replaced by many closed-ended ones. These questions were then further refined and purified using various statistical procedures and eventually whittled down to a handful. A prime example of this is the measurement of intelligence. Early intelligence tests required answers to hundreds of questions. Now intelligence can be reliably measured by the Cognitive Reflection Test using just three questions.[3]

Having so few questions makes it possible to capture the thoughts of many more people. Closed-ended self-reports also permit the use of numerical rating scales, answers to which can be statistically analyzed. This can be done on a real-time basis using survey software, making it possible to study large phenomena quickly and efficiently. That is why public opinion polls and a large swath of academic social science research relies on such questions.

Of course, not everything can be captured using structured and closed-ended response scales. Some things that happen deep within the bodies and minds of people are far too nuanced for that, so open-ended questions are still relied on. Take, for instance, cognitive research on how people process visual information in their brain. This body of work led to the understanding that people apply heuristic thumb rules or systematically process information (which is part of the SCAM's explanatory framework). The stream of empirical research that led to this understanding was based entirely on open-ended responses.

In a typical cognitive processing research study, participants are shown persuasive messages from a variety of sources (e.g., a news article from CNN and from an unknown blog). Individual participants are then asked to recall whatever they read. Their self-reported, open-ended responses are then analyzed using the same content analytic approach that was used to develop the V-Triad. Human coders read each response and assess whether the individual used a heuristic or systematically processed the article.

While cognitive psychology has long used open-ended questions in this manner, security research on users seldom utilizes it. Part of it stems from the engineering background of researchers. Most just aren't trained in empirical social science. Even when security researchers include open-ended questions in their surveys, they tend to ignore the responses or use them to strengthen the conclusions they have already arrived at using closed-ended measures.

A prime example of this is the MITRE research on user training I discussed in chapter 4.[4] As you'll recall, their research team conducted a series of experiments to examine the value of training and found that awareness training was largely ineffective. This conclusion was based on behavioral and self-report data from a variety of closed-ended questions. Their research, however, also included an open-ended question that invited users to comment on why they fell for the phishing pen test. One respondent stated, "I clicked on it inadvertently without thinking." Another said, "I just got the email and clicked on the link. A webpage came up, but it seems suspicious." The MITRE team used these responses to explain away potential contaminants in their research design, but not for drawing conclusions about users. For that, they relied solely on their closed-ended responses.

Of course, the other reason they ignored the open-ended responses is that MITRE's team lacked a framework in which to interpret them. If the team had used the SCAM, they would have quickly noted the influence of habitual automaticity in the first response and the play of cyber risk beliefs in the second. In fact, the user even spelled out the influence of suspicion. If the researchers had used the SCAM framework, the closed-ended questions and open-ended responses could have complemented each other. Closed-ended questions could have captured whether suspicion was triggered, and answers to open-ended questions could have explained the mechanism. Incorporating both types of questions would have captured how users were deceived quantitatively and explained why qualitatively. In other words, it would have captured the breadth and depth of the process of user deception.

This example demonstrates the value of closed- and open-ended questions. Closed-ended questions allow quantification, and open-ended questions get richness in response. Closed-ended questions permit high-level statistical analysis. Open-ended ones allow granular understanding of the user; they need to be quantified to allow statistical analysis. In practice, because of the little time it takes for users to answer, a survey can include many more closed-ended questions, while open-ended questions need to be limited because of the time required to respond to them. They complement each other when their design is guided by a comprehensive explanatory framework. When this is done, they can provide deep insights into human thought and action.

Using both types of questions is necessary when dealing with mind-level phenomena such as deception or its detection. In such research, actions, or what people do, is rather easy to capture using clickthrough tracking. It's the mind-level processes—what users thought that led to their actions, the thoughts that forestalled their actions, and the actions that occurred without thinking—that researchers need to capture. Behavioral tracking alone can never get to the reasons. This requires self-report questions, ideally those that can capture the deep mind-level thoughts that can be compared to action, in both its breadth and depth. It requires a combination of closed-ended and open-ended questions designed using the SCAM framework.

### THE DIAGNOSTIC

The initial work on the SCAM used dozens of closed-ended self-report questions. There were questions measuring suspicion, risk beliefs, heuristic processing, systematic processing, habits, and self-regulation. With the addition of demographic questions, a survey would have over 50–75 questions. This made the approach impractical for repeated use in organizations. On the plus side, the approach gathered a lot of quantitative data that could be analyzed statistically, but the downside was that interpreting the data required knowledge of regression modeling and social science data analysis. Most organizations just weren't equipped for that. Not only did they not want to subject users to repeated lengthy surveys, but the results were also difficult to communicate to their users. This meant that the findings, although insightful for policy-making, couldn't be easily communicated to users. This also meant that the users couldn't be held accountable for something that was incomprehensible.

A simpler approach was needed. It had to be practical, so it could be used repeatedly. It had to be easy for users to respond to, so organizations could gather adequate responses. It also had to be easy to interpret, so both IT staff and users could comprehend the results and act on them.

To achieve these objectives, my research team worked on a scale purification process. This is an iterative process where different variations of the core questions from the SCAM were introduced into the pen tests I conducted in different organizations. By contrasting how well each question predicted individual phishing deception or its detection, the relative contribution of each was examined. This iterative process helped remove the chaff and identify the "best-fitting" questions—those that predicted the most, introduced the least amount of bias, and were the easiest to implement using a survey. One question among all consistently stood out, capturing the breadth of users' phishing susceptibility. This single question was most indicative of everything the user considered or did leading up to the attack's outcome. This one question measured suspicion. It was a closed-ended question and could be followed up with an open-ended question no different from the one used by MITRE. Together, these two questions could replace dozens of others. Best of all, they could be implemented following a pen test in an organization, no matter its size, location, or user population. These questions were also simple to understand and could be quickly answered by most users. In other words, in mere seconds they could help establish the quality of a phishing test and the likelihood of someone detecting it or getting deceived by it. The closed-ended question is shown in figure 6.2.

Respondents could choose any number from 0 to 10 to indicate the level to which a pen test triggered their suspicion. While this question quantitatively measured the user's vulnerability to any attack, a follow-up question could measure their reasons. This could be simply accomplished with an open-ended question (figure 6.3).

Together, the two questions capture users' susceptibility to phishing on quantitative and qualitative levels. They measure the facets of the users' thoughts and actions, from all the angles that earlier necessitated dozens of questions. They work best when posed after a phishing pen test and when the user responses are analyzed using the SCAM framework. The overall process, which begins with the pen test and culminates with the analysis and diagnosis of user risk, is called the Cyber Risk Survey (CRS).

The CRS begins with a phishing pen test but not just any test. It is an ideal test, one that follows the V-Triad for its design. The test includes elements from the vertices of the triad and is validated before deployment. Validation is done by a small subgroup of respondents, either the IT staff who crafted the pen test or a few users who are not part of the routine pen

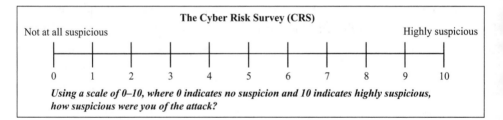

**Figure 6.2**
The closed-ended question from the Cyber Risk Survey.

*Follow-up question:*

*Using complete sentences, please explain in detail the reasons for this level of suspicion.*

**Figure 6.3**
The open-ended question from the Cyber Risk Survey.

**The Cyber Risk Survey (CRS) for establishing baseline**

Not at all suspicious — Highly suspicious

0 1 2 3 4 5 6 7 8 9 10

***Using a scale of 0–10, where 0 indicates no suspicion and 10 indicates highly suspicious, how suspicious is any average user in the organization going to be to this attack?***

***Follow-up question:***

***Using complete sentences, please explain in detail the reasons for this level of suspicion.***

**Figure 6.4**

The Cyber Risk Survey for establishing baseline.

test. They serve as proxies, assessing the perceived potency of the pen test they are given, and help establish a baseline of its likely impact. They respond to the very same CRS, with a slight rewording (figure 6.4).

The responses to the baseline estimation questions along with the responses from users subjected to the pen test are taken as a whole to assess individual cyber risk and organizational readiness. The specifics of how all this works together are presented in chapter 7.

At this point in the book, you are probably wondering whether the V-Triad and the two-question survey approach even work. How do we know the triad is appropriate? Is there precedence for it or for its survey approach? Well, here's triangulated evidence for it.

## TRIANGULATION

Triangulation is a simple idea but with profound implications. At the most basic level, it involves using different approaches to study a problem and assess its validity. The idea is that if different scientists, using different approaches and different data, come to the same conclusion, then the conclusion is more than likely valid. It is even better when the scientists are from different fields and are working independently of each other. This way, paradigms, as in the lock-in from being within a field where everyone thinks similarly, are less likely to influence the work.

Triangulation of evidence is the gold standard for many human decisions. It's how juries work. It is also how many medical diagnoses are done. A physician might conduct an exam in the office and order different tests. The tests come with different conclusions from radiologists and pathologists. The combination of these expert insights allows an effective diagnosis.

We can apply a similar standard to assess the validity of the V-Triad and the CRS measurement approach. As I mentioned, the models for the SCAM were developed around 2015, and the V-Triad was developed in 2016. In 2020, a team from the NIST developed what they termed a Phish Scale.[5] The scale was built to improve the quality of phishing pen tests for all the reasons the V-Triad was developed. The approach the NIST team took was also identical. They examined different phishing emails and correlated their features with their clickthrough rates. Of course, they used a different corpus of pen tests, but nevertheless their conclusions were more or less identical. In place of the three vertices of the V-Triad, they found two: cues in the email and their compatibility or match with user routines.

But there is more. Even the implementation of their scale is similar. They ask IT staff to list the cues in emails and, using a series of closed-ended questions, rate their perceived difficulty and overall fit. So the NIST team approaches the same problem, uses the same methods, and comes to similar conclusions. Because they worked with a different corpus of emails and worked independently, the concurrence of their findings helps triangulate the research. It helps firmly establish the validity of the V-Triad.

There are differences, however, that show why the V-Triad is superior for developing the ideal pen test. For one thing, the Phish Scale takes a very broad view of cues. According to the NIST, a cue is anything that has a visual anchor in the attack. This conceptualization of cues is inconsistent with the principles of cognitive science, where cues are clearly defined and distinguished from textual content.

This is important because while using the Phish Scale, the IT department has to list each cue. This means identifying anything on the surface from text (display name and subject line to persuasive appeal) to graphics (attachment indicator, brand name). As we discussed in chapter 4, online communication surfaces are rife with graphics, text, and icons. Using the

NIST approach requires listing a plethora of cues and then individually rating each of them. This has to be done not once but twice: first for their perceived difficulty and then for their fit.

This is very difficult in practice because it is hard to establish why something is a cue and when it isn't. For instance, is a form field for entering credentials a cue? How about the profile photograph of a sender that some email apps autopopulate? It's surely a cue when it is present, but what if it is absent? Is that not a cue if users expect it?

The problem gets compounded when dealing with textual elements in emails, which are also inaccurately treated as cues by the NIST approach. As close to a century of cognitive research has shown, textual appeals have a different cognitive effect compared to cues. As you'll recall, the former triggers systematic processing, the latter heuristics. It's why the SCAM, which forms the basis for the V-Triad's explanatory framework, distinguishes between different cognitive processing styles.

Conflating cues with text, however, poses yet another problem. An email's text often contains multiple cues, sometimes in a single sentence. A single sentence can have a warning, a deadline, and grammatical mistakes. Are these counted as one cue or three? What about grammar versus spelling mistakes? Would that be one or two?

Persuasive phrases and appeals complicate this even more. Let's take as an example a phishing email appearing to come from the American Red Cross that asks for donations with a deadline and few typos, even a warning to be cautious about phishing (which many attackers introduce to enhance trust). Using the NIST approach here means either ignoring some cues (such as ignoring the appeal in the Red Cross phishing email and focusing on the deadline, the name of the American Red Cross, and typos) or counting the appeal and all the elements in it—which would double or triple count their impact. This may vary based on who is doing the coding, adding more contaminants into the data. Thus, the broad view of text as cues leads to a highly unreliable rating system.

Finally, the Phish Scale ignores customizability—an important factor that phishing preys on and the V-Triad includes. As discussed earlier in this chapter, customizability captures the facets of phishing attacks that take advantage of the mental accommodation users make in online environments. As every one of us can attest, a large part of what we do online involves transferring knowledge from one surface experience to another. We have to do this because of the constantly updating operating environments and user interfaces. And we don't use instruction manuals for any of it. Ignoring this pivotal facet of cyber user behavior misses a fundamental aspect of what makes users susceptible to attack. Thus, the Phish Scale takes a limited view of users as being solely reactive to cues rather than being adaptable.

Notwithstanding these limitations, the Phish Scale finds a strong relationship between an attack's credibility and compatibility and the actual phishing click rate. Overall, it supports the V-Triad and finds evidence to support its dimensions. The triad approach is far more comprehensive and easier to implement. Rather than counting cues and repeatedly rating each dimension's intricate constituents, it gets directly to the heart of the issue—how much suspicion the deceptive elements of a phishing email individually or altogether arouse. After all, it is the totality of factors in the email that work in tandem and deceive. The CRS also pools the estimates of IT staff. This captures the consensus view of expert users in the organization, which helps reduce variability and bias. Thus, the approach accounts for the cognitive, behavioral, and experiential factors in users that have scientifically been shown to predict deception via phishing or its detection. Now that you know how well it works, in chapter 7 we can put it all together and conduct a cyber risk assessment.

# 7 CONDUCTING A USER CYBER RISK ASSESSMENT

Why conduct a phishing pen test? If you have followed the rationale in this book so far, you know it's not for training users. While a pen test could help in understanding who among the users may need training, it doesn't train them, especially not when it follows a "shock and awe" approach—designing pen tests solely to trick users, who are then punished for being duped. Instead, we conduct pen tests to understand who among our users is at risk. We do it to quantify their levels of risk, diagnose the reasons for their risk, and identify what we can do to protect them. We do it to understand who needs training and what kind or who may need something other than training. We do it to identify the weakest links among our users and protect them. A phishing pen test developed and implemented using the Cyber Risk Survey (CRS) can do all that.

But there are other, higher-order reasons as well. Pen tests aren't meant only to identify the cyber risk one user poses. They need to help assess the severity of the threat of social engineering to the overall organization or enterprise. The test also needs to quantify the organization's risk posture. It further helps if the quantified assessment is scalable to the enterprise or even an entire sector. This way, a policy maker or insurer can compare risks at different levels. They can then mitigate the threat of phishing by implementing safeguards, allocating capital, and deploying resources where necessary. They can also efficiently allocate training, track the efficacy of various mitigation efforts, and better deploy capital and resources.

Using the CRS, IT departments can also compute a net resilience score (NRS), which quantifies risk and readiness at the enterprise level. Derived from the CRS, the NRS is a pooled estimate that quantifies the cyber risk from phishing in a group of users across one or a series of phishing pen tests. The NRS has a maximum value of 100 and is scalable upward to an organizational division or even the entire organization. This makes it possible to compare the relative risk among organizations that implement it. It can be scaled up so the overall risk across a sector, a region, or across nations can be understood. Thus, the widespread use of the CRS and its associated metrics can help in strategic planning and policy-making at the micro-organizational and macro levels. This chapter explains how the CRS does all this.

## MEET THE DOCTOR

To begin, let's extend an approach we are all familiar with—a patient's visit to the physician. Physicians the world over follow these broad steps during such encounters:

- a patient comes with an underlying ailment;
- the physician conducts procedures to assess or trigger the symptoms of the ailment under controlled conditions;
- the physician uses a diagnostic or test to measure the patient's response to the procedure;
- the diagnostic has a baseline or standard of performance;
- the patient's performance is compared against this baseline to assess the ailment;
- based on the assessment, the patient is diagnosed and a treatment or solution is derived;
- the physician communicates the diagnosis to the patient.

This fundamental approach to medical diagnosis is also the basis for conducting user cyber risk assessments (figure 7.1).

### Step 1: Users Suffer from Cyber Vulnerability

The fundamental assumption underpinning user cyber risk assessments is that all users are vulnerable to social engineering. Another is that this vulnerability varies among users. While it may seem obvious to state this, it bears noting that this isn't the assumption of today's embedded training approaches.

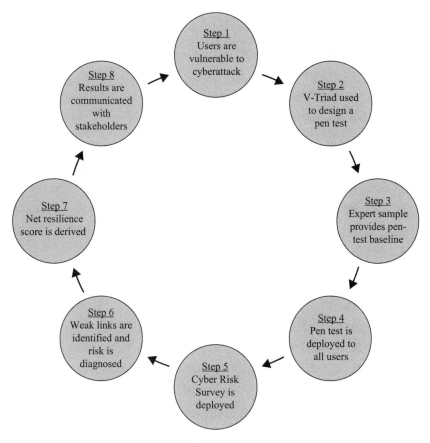

**Figure 7.1**

Steps to conducting a user cyber risk assessment.

Most test only a subset of users (usually ignoring the IT department and top management), use a pass-fail measure, and focus only on those who fail. These approaches forget that everyone is a target of social engineering, even those in the IT department and top management, who are often sought-after targets because of their access to core data. Today's embedded training approaches also ignore everyone who passes the phishing test and focuses only on those who fail, and they intrinsically assume that users either pass or fail and that there is no real variance or degree in between (e.g., inadvertently passing or failing).

The CRS begins with the correct assumption that everyone is at risk from social engineering; that everyone should be subjected to testing, including those in the IT department and top management; that pen testing reveals more than who fails; that the point of pen testing isn't to ignore those who pass or shock those who fail; and that there are a variety of reasons for both. The CRS is fundamentally focused on finding and capturing all reasons for user vulnerability from phishing in the most valid and reliable manner.

### Step 2: IT Managers Design a Spear Phishing Penetration Test

The CRS requires that IT staff craft an ideal phishing pen test (or a series of them) using the V-Triad. It is necessary to incorporate one major element representing each vertex of the triad—credibility, customizability, and compatibility. Two main considerations guide the choice of specific elements from within each vertex. First, each element should contribute toward the email's ecological validity. It should help make the test as realistic as possible. So, a pen test mimicking an iTunes email may contain an Apple logo and use its font and colors, but the goal isn't to keep adding unrelated icons and form fields to address each vertex. All additions need to pertain to the context; otherwise, the resulting data would be unreliable. Second, the pen test's quality must be known. Its ability to deceive must be quantified prior to its deployment. The way to achieve this is through an empirical examination, which is accomplished in step 3.

### Step 3: Establish a Baseline for the Pen Test

After design, the IT department needs to determine the pen test's quality. This can be accomplished using a small subset of representative users from the organization or even with a few IT staff who can serve as an "expert" sample. The baseline is established by asking each user in the subsample to rate the pen test independently. Each answers the two CRS baseline questions measuring how suspicious the average user in the organization would be and the reasons for this rating.

The validity of the baseline score comes from the integrity of the users who are acting as proxies for other users. If there is any concern about how well they represent the test's targets, a simple fix is to ask each of them to rate their similarity to the users in the organization. A question such as this is all it takes: *Q. Using a scale of 0–10, where 0 indicates "completely similar" and 10 indicates "completely different," how similar are you to the average user in the organization?* In general, anyone who scores 8 or more on this scale is considered too dissimilar to the average user, and their responses are excluded from the baseline.

Baseline scores can be generated with as few as one IT staff member serving as an expert in a smaller organization and with as many as 25 users in larger organizations. Large samples aren't necessary unless the organization's users vary considerably. What is crucial, however, is that the individuals who serve as experts be clearly informed that their task is to assess what *an average user* in the organization is likely to do, not what they themselves would do. It is also important that the experts perform the assessment independently, without conferring with each other. The baseline score can be computed prior to the deployment of a pen test or for picking a test from among different exemplars that are being considered. It can also be assessed simultaneously when the pen test is deployed to all users in the organization.

The responses provided by the subset of experts are then averaged to come up with a benchmark expectation from the pen test. For instance, if the average across the subset for a test is 6, then the conclusion is that to be considered low risk most users in the organization should ideally achieve a score of more than 6 on that pen test.

### Step 4: The Pen Test Is Deployed on the Organization's Users

Next, using email marketing software or pen-testing software, the phishing email is deployed to users in the organization. Pen-test applications and email marketing software allow tracking when any user opens the email and what they click on. This provides a behavioral measure that can be used as a cross-reference to check the veracity of their survey responses. If IT staff do not have access to pen-testing software, an email can be sent using any standard whitelisted email account, with a clear GIF tracking cookie embedded in it. All this can be achieved using freely available email-tracking tools.

Regardless of how the pen test is ultimately deployed, it is imperative that there be an accurate list of the users who are targeted. In many large organizations, there are email accounts of employees who have left the organization. In some cases, users have multiple email accounts and even use personal email accounts for work purposes. It is important to ensure that the email to which the attack is sent is valid and that the number of times any user is targeted is accurately counted.

### Step 5: The Cyber Risk Survey Is Deployed

Following the pen-test deployment, the two-question CRS is deployed. Because the risk diagnosis is based on absolute and comparative data, the survey needs to be sent to everyone, including those who fell victim to it (those who opened the email and clicked on the hyperlink or attachment) and those who detected the deception.

The CRS can be deployed in one of two ways. It can be embedded in the hyperlink within the phishing test. In this case, rather than send a phony phishing hyperlink in the pen test, the link can instead be to the survey, which opens whenever the user clicks and falls victim to the test. Those who don't fall for the phishing test can then be sent the CRS within a week after the attack, with an image of the phishing pen test email in the survey to aid recall. Alternatively, all users, those who detected the threat and those who were deceived, can be sent the CRS a week after the pen test. Both approaches net comparable data, and in my tests I have found limited variance in recall between the immediate- to seven-day window.

The overall goal is to have at least 60 percent of all users who were targeted by the attack respond to the CRS, a response rate necessary for ensuring the validity of the data. To ensure this response rate, IT managers must make multiple attempts to procure responses from everyone targeted.

### Step 6: Weak Links Are Identified, and the Reasons
### for Their Vulnerability Are Diagnosed

Using the CRS, the user's performance on the pen test is compared against the baseline numerical cutoff. Following the earlier example where we netted a baseline of 6, any user scoring less than or equal to 6 would be considered relatively high risk. These users are the weak links, and among them the users scoring the lowest are the weakest link(s). Anyone exceeding the threshold, in our example scoring 7 or more, would be considered lower risk (see figure 7.2).

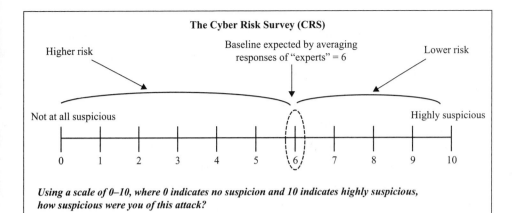

**Figure 7.2**
Scoring the CRS.

The open-ended responses of those scoring 7 and over are then examined to diagnose the reasons why they became suspicious. For this, IT staff applies the SCAM and quantitatively codes the users' open-ended responses. I demonstrated how this is done in chapter 4. Following the same approach, IT staff can review each user's response for indications that they used a heuristic, a systematic or elaborate thought, a cyber risk belief (CRB), or acted habitually and nonconsciously. They can even count the number of instances of each, which can then be averaged. Following this coding protocol, IT staff would be able to estimate the average CRBs, systematic processing, heuristic processing, and habit activations required to detect deception. This can be treated as the desired threshold value of each parameter required to detect deception.

The computed average value of high-risk users' CRBs, systematic processing, heuristic processing, and habit activations can then be compared against the threshold to identify gaps. For instance, if on average there are four systematic thoughts applied by low-risk users, each individual high-risk user could be examined to see how many systematic thoughts they applied. The goal is to see whether they met or exceeded the standard. The same examination could be conducted to assess the relative difference in CRBs, systematic thoughts, and habitual reactions of users who are high risk versus all the users who were low risk. This relative difference helps diagnose each user's lack and for understanding why they are at risk from social engineering. Based on this information, users can be accorded more training, different types of training, or other technical solutions.

### Step 7: Net Resilience Score for the Organization Is Derived

The net resilience score (NRS; see figure 7.3) is the percentage of users who are low risk subtracted from the percentage of users who are high risk. The metric's value theoretically ranges from −100 to 100, with a lower number indicating increased vulnerability and a higher number indicating greater resilience. For instance, if 70 percent of the users in the organization were judged as high risk based on the CRS and 30 percent were judged as lower risk, the resulting NRS would be 40 (i.e., 70–30).

The NRS can be computed after each pen test or by averaging the high and low risk percentages across tests conducted over time. The score is a derivative of the strongest causal predictor of phishing deception that also accounts for the difficulty level of phishing attacks. An NRS of 100 would indicate that the organization is completely resilient against phishing. Conversely, an NRS of 0 would suggest that the organization is not resilient, while a negative number would indicate high vulnerability.

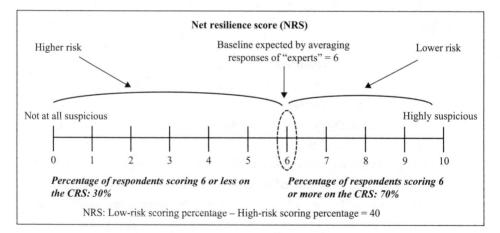

**Figure 7.3**
The net resilience score.

### Step 8: Results Are Communicated

Once derived, the NRS along with the diagnosis of risk needs to be shared with users in the organization, not just with senior management. Often we skim over this requirement, but this is more important than it appears. To achieve cyber resilience, we need buy-in from everyone, especially the users who are the weakest links and pose the greatest risk to the enterprise. Going back to the purpose of a pen test, with which I started the chapter, it isn't to test people or individuate and identify them. It is to cure them of their risk.

Communicating results is pivotal in this process. It lets users know why they are at risk, what they did wrong, and what others in their social group did right that they, too, could do to reduce their risk. The process of cyber risk assessment is incomplete without communication. At this stage, IT managers can decide whether they need more tests, what those tests might assess (such as target user habits vs. risk beliefs or heuristics), whether the test's difficulty needs to be enhanced, and who should be targeted by the next test. Then we'd restart the process from step 1.

In summary, the CRS-based process for diagnosing the weak links in the organization follows these steps:

- begin with the assumptions that all users are vulnerable;
- develop a valid spear phishing pen test using the V-Triad framework;
- ascertain the baseline performance expectation of the phishing pen test;
- deploy the pen test on the users of interest;
- use the CRS to measure the responses to the pen test;
- compare users' performance on the pen test versus the baseline, identifying the low-risk and high-risk users;
- diagnose the reasons for high risk among users by comparing low-risk users' open-ended responses with the high-risk users' responses;
- compute the NRS by deducting the percentage of low-risk users from the percentage of high-risk users on a per-test basis and aggregate it on an ongoing basis.

Once the weak links are diagnosed, there are a number of interventions and solutions that can be applied. Many of these are discussed in chapter 9. For now, let's stay with the CRS and discuss why it works and why it is better than anything we are presently doing in cybersecurity.

## THE ADVANTAGES OF THE CRS APPROACH

The CRS along with a pen test designed using the V-Triad captures the breadth and depth of processes that lead to social engineering victimization. The approach is vastly superior to the existing approaches for the following reasons.

First, as discussed earlier, almost all spear phishing tests are ad hoc, and there is no basis for designing them. They range in sophistication, quality, content, and structure, making them unreliable. In contrast, each element in the CRS has a basis in the science of security.

Second, no mechanism presently exists for ascertaining the quality of a pen test. Because of this, it is impossible to know whether a pen test is valid or even applicable for testing. The CRS approach, however, leads to a pen test whose merits can be quantitatively ascertained, making it possible to know whether an individual test is suitable for use in an organization and the extent to which it is comparable to other tests.

Third, the CRS approach of assessing the quality of a pen test can be done a priori, before it is deployed on actual users. This means a pen test can be corrected if it doesn't meet the right benchmarks and not used. Moreover, the quality of the test can be quantitatively ascertained so we know how comparable a test is across its implementations.

Fourth, all social engineering pen tests focus on pass-fail results. None provides any insights into why any user fell or didn't fall victim to the test or how well they did compared to other users. The CRS enables us to understand the degree to which users are susceptible to deception. Data from the CRS provides a quantitative estimation of who is at risk, helping pinpoint the weakest links among users. It also captures the depth of reasons at a granular level, thereby explicating the underlying reasons for each user's risk or resilience.

Sixth, in all pen tests, only those who fall victim are counted and treated, usually with even more training. The underlying presumption is that only the victims of a test are susceptible to phishing. The tests also assume that everyone who didn't fall for it did so because they were reasoned and cautious. The CRS makes no a priori assumptions about who needs to be treated. It locates their risk or resilience in their cognitive, behavioral, and personality-based triggers. It gets into the mind of the user and provides a risk estimation based on how they think and act, and thereby comprehensively explains why someone fell for the phish or detected it.

Seventh, because of the lack of a framework outside the technology used to conduct a pen test, all training protocols that implement pen testing find solutions in technology. Many are conducted by IT staff with an engineering orientation, who, extending their mechanistic assumptions about users, end up either advocating more training or asking for further technical protections. The testing data lacks the granularity and richness needed to understand or diagnose the reasons for each user's risk.

In contrast, the cognitive-behavioral framework of the SCAM underpins the CRS. This helps uncover issues that might be unique to users or that are present in certain organizations. For instance, there could be a unique heuristic, a flawed cyber risk belief, or a quirky ritual that is endemic to an organization or user type. The CRS allows IT staff to discover and fix these. In many cases, this can lead to solutions that don't even require more training. (Examples of this are presented in chapter 9.)

Eighth, the CRS can provide a robust estimate of user cyber risk with fewer pen tests, unlike the current practice, where organizations repeatedly conduct many pen tests. The CRS makes this possible because its tests are based on the V-Triad, which is built on the SCAM's scientific framework. Because of this, IT staff can create a few reliable tests, each with known levels of difficulty, and implement just them rather than having to create many tests to adjust for their unreliability. The tests could also be interspersed throughout the year, followed by various interventions, whose efficacy could be judged using another application of the CRS. This would help identify the impact of each intervention and why it succeeded or failed.

Ninth, the results of the CRS help compute an NRS for the overall organization, which provides a snapshot of its overall resilience against social engineering. Being easy to compute and understand, the metric can be shared with other organizations. They can share data on the tests they used, the interventions they applied, and the changes in the NRS over the course of these applications. This can help organizations understand their relative levels of readiness and locate solutions that are workable, without having to reinvent them each time.

Finally, the CRS and its associated metrics can be easily communicated to users. Today, any such communication is done with users who fail a pen test. It is done to create fear, not build accountability. Part of the reason is that the testing paradigm uses a shock and awe approach—reinforcing users by triggering accidents. In this view, IT managers are like police officers, looking to give out tickets and harsh sentences rather than encourage better behaviors. Even developments such as the Phish Scale are for IT staff. They are not tools designed for communication but are metrics for better policing.

The CRS inverts this. It fosters communication by simplifying the outcomes. Anyone with limited statistical knowledge can respond to the CRS, implement it, derive an NRS score, and interpret it. This makes the approach easy to implement and ideal for sharing and communicating.

Once they know their scores, users can understand where they stand in relation to others in their organization. They can understand what they did right, what they didn't, and what they need to do differently. They can even understand the IT department's expectations and the reasons for them. In this way, the CRS fosters transparency and shared understanding. By using the CRS approach, the goal of IT staff shifts paradigmatically. They are no longer policemen but instead are change agents who work to understand users and reduce their risk from phishing.

## LIKE BAKING A CAKE

It is certainly simple to bake a cake. Take flour, eggs, butter, salt, sugar, and baking powder, add milk, mix, and heat until done. These seven ingredients could be turned into a masterpiece in the hands of an experienced pastry chef and a disaster in those of a novice. What differentiates them is expertise that comes from years of trial and error. Such trials for a novice could go on in perpetuity because anything from baking cakes to writing music can be done in a multitude of ways. The expert instead knows what works from experience and can craft a great product in fewer attempts.

This is the case with the application of the CRS as well. It is simple but can be applied in a multitude of ways, leading to all manner of different outcomes. Thankfully, over the years, the SCAM studies and the CRS applications have already led to an understanding of what does and doesn't work. You, too, can learn from the trials and errors of those who have already implemented and addressed its blind spots. This way, you don't have to drive down different dead ends. You can quickly apply the CRS and get the cyber risk estimate you desire. The following are answers to the most frequently asked questions that arise when implementing the CRS.

Q: Do two questions suffice? Are there other examples of few questions netting valid results?

A: Most polls use a single measure when assessing public opinion. Even intelligence tests, such as the Cognitive Reflection Test (CRT), discussed in chapter 6, provide valid estimates using just three questions. In addition, there is also the Net Promoter Score[1] developed by Bain & Co., which is widely used by Fortune 500 companies for making strategic decisions. It, too, uses a single question.

Q: Do people in IT departments respond to surveys and analysis of open-ended responses? Where is the precedent?

A: Surveys have been used for a while in IT departments. The SANS cyber security model employs closed-ended questions that measure user attitudes. Even the Phish Scale requires that IT staff answer questions about phishing email features. Not only are users accustomed to answering such questions, but IT departments are also used to analyzing such responses.

Q: I want more quantitative data and would like to use quantitative, closed-ended questions instead of the open-ended responses. How could I do that?

A: The closed-ended questions are available in the research that presented the SCAM model. There is an updated version that uses 20 questions, which can be used in a multivariate regression (e.g., using the Lens Model) to measure individual risk predictors. Keep in mind that the use of these models doesn't change the risk profile that we determine using the CRS approach. All it does is provide a statistical model of its predictors. We can achieve the same result using the numerical codes given to the users' open-ended responses in the CRS. They, too, provide a quantification of the cognitive-behavioral factors influencing suspicion.

Q: What about the other measures we already have from pen-testing software?

A: For organizations using a pen-test product, there are several measures such software provides. All of them can be incorporated into the CRS data. For example, if the software provides data such as the email open rates (as in who has opened the pen-test email), it can be correlated with the risk ratings from the CRS and for cross-referencing the responses. This can, for instance, tell you whether users are risky because of how quickly they respond to an email. You could then create interventions aimed at reducing their habit scores.

Q: Should everyone be assessed using the CRS?

A: Unlike the current practice of tracking and training only those who fall for a phish or counting failure, the CRS assesses everyone in the target groups of users.

Q: What about measuring user personality?

A: My research, as well as that of others in the field, has examined the correlations between personality and the factors relevant to deception detection. Decades of research have found no relationship between the personality of people and their mode of cognitive processing. The only effect of personality is on the strength of habits, and that is already accounted for in the CRS framework.

Q: What about demographics? Should I measure age, gender, education, position, and others?

A: Measuring a lot of user-level factors simply for the sake of it is not only unnecessary but will also increase the length of the survey, affecting its overall completion rate. Having tested for the influence of demographics on users' vulnerability to social engineering, I can say with confidence that none has effects that are stable enough to warrant inclusion. Some factors, such as a user's age, matter but only at the extremes. That is, if you have an employee who is significantly older than the rest, and if that individual is largely untrained in using relevant technology, it might affect their vulnerability to social engineering. Other than this, education, income, race, and gender don't matter.

The only other factor that has been more impactful is length of tenure in the organization. New employees are far more likely to fall prey to attacks that most others in the organization who have been enculturated into the organization's norms and communication patterns would detect. So, an attack that mimics a file-sharing service that is used within an organization is more likely to net relatively newer employees, who might ignore some obvious patterns in the email that those with more experience would clearly catch, but we already know this. We can, however, account for this through our pen-test benchmarking process by inviting a few new users to rate the pen test's quality.

Q: How many users do I need for a reliable benchmark score?

A: The consensus CRS benchmark can be procured using a handful of users. In small organizations, even a single valid user can provide a benchmark. In a midsized to large organization, ideally, anywhere from 5 to 25 users or even IT staff members should provide the CRS benchmark. If there are any concerns about the quality of the experts in the sample, all you need to do is measure each expert's similarity to the average user (as I presented earlier). You can use the response to exclude anyone who isn't similar to the average user in the organization.

Q: Do I need to get buy-in from top executives?

A: Getting buy-in from top leadership in the organization is very important. In my opinion, it should be done before conducting any assessment of user risk. It is also important that the top leadership communicate their support directly to users in the organization. This improves participation and response rates. If you are an external consultant to the organization, ask the chief information security officer or someone from top management to solicit participation. Users are far more willing to participate when the instructions come from within the organization and from those they respect.

Q: How do you ensure that users participate and that they respond to surveys honestly?

A: You get users to respond by first ensuring that they recognize the value of participating in the CRS. The value of this needs to be communicated, and it works even better when such communication comes from people in the organization who users trust and respect. It is also important that IT staff approach every pen test as an opportunity to learn about users, not trick them or show them to be less adept. This requires encouraging participation rather than demanding it. Finally, ensuring participation requires communicating the results of the CRS in a manner that improves the users' knowledge and investment in the organization's cyber readiness. This ushers in a culture of shared cybersecurity responsibility and helps improve the organization's cyber posture.

Q: How many responses to the CRS do you need?

A: Ideally, you need to procure responses from at least 6 out of 10 users who were subjected to the pen test. This is not a blanket requirement, and the data must be assessed to ensure that the 60 percent who respond aren't mostly from one group (i.e., just those users who didn't fall for the attack). In general, achieving such a rate is not difficult as long as you (1) procure buy-in from leadership, (2) inform users about the organizational support for the CRS, (3) communicate the value of the CRS, (4) conduct the CRS only a few times each year, (5) communicate the CRS and its results to the users, and (6) build a culture that fosters communication, transparency, and participation.

Q: How long should I pursue responses or keep asking users to complete the CRS?

A: It is good practice to send at least two reminders to all users who have yet to respond to the CRS in the week after their responses were first solicited. If you deploy the pen test on Monday and send the CRS on Friday, a reminder should be sent the following Tuesday and then another on Friday. Reminders should be sent only to users who have yet to respond. In my experience, reminders are far more effective in procuring participation when they come from the leadership in the organization.

Q: Should I communicate the results of the assessment to the end user?

A: Most IT managers seldom inform users about their individual test results. In current awareness training protocols, IT staff members tend to communicate only with those who fail the test and need remedial training. The CRS is a tool for information acquisition from users and is designed to be simple enough for the information to be shared. Sharing the results with each user is also necessary to encourage participation in future applications of the CRS. As a case study presented in chapter 9 will demonstrate, the open-ended responses from the CRS can provide rich insights that can inform best practices that are more easily accepted by users because they come from other users. All this begins with the communication of the results of the CRS. Communicating the CRS results also helps users reflect on the inherent risks of their online thoughts and actions, leading to more careful actions over time.

Q: Where can I find pen-test exemplars, and should I download them online?

A: You could create your own or reuse ones that you find online, but the general rule is to ensure that they meet the V-Triad's minimum requirements. Regardless of how you develop your test, you must ensure that you assess its baseline value using the CRS. This ensures the test's overall fit. You can also create multiple pen tests and assess all of them simultaneously. This helps develop tests that increase in difficulty. It also helps ensure that the users in the expert subset don't know which of the exemplars are going to be used in the final pen test. This way, they cannot inform other users about which test is forthcoming.

Q: Could I reuse the same pen test already deployed?

A: There are two ways to reuse the same attack. One is to reuse it within the same test episode. This would be deploying the same test, say, on Monday and then again on Friday. I have done this in the past and found that there are those who click on the same attack multiple times—an issue that cannot be captured in a single-episode pen test. The other way to reuse a test is to conduct the same attack a few months later. In such cases, the baseline survey needs to be repeated to account for any increased awareness of the attack.

Q: How many assessments per year should I conduct using the CRS, and how far apart should they be?

A: The surveys should be conducted a few months apart, especially if there are interventions that are being applied between them. This way, the data can capture the gains or losses in them. Most organizations should repeat the approach three to four times a year—that is, once every quarter—in order to be able to capture the changing routines of users.

Q: How many pen tests should be given per year, and how far apart should each test be?

A: While some organizations conduct pen tests throughout the year, giving too many tests actually reduces the quality and accuracy of the tests. They end up training users into expecting the test and knowing when a test is coming rather than learning to spot spear phishing. Instead, the focus should be on conducting fewer tests of better quality. This can be done by deploying a test every quarter, each increasing in difficulty and calibrated to capture the improvement of users.

Q: What are some ways to ensure response quality?

A: Ensure that the instructions on the survey are clear. Respondents must be asked to be honest in their responses and provide complete sentences when explaining why they were or weren't suspicious of the email. It helps to procure buy-in in advance. This can be done by communicating the value of the CRS, explaining to users what it entails, and laying out how they could personally benefit from responding to it. These benefits could be anything from having to do less repeated training to implementing solutions that are personalized to their security needs so they have less to worry about when it comes to cybersafety.

Q: If the difficulty of the test increases, should I adjust for that in my analysis? How do I do that considering that the closed-ended 0–10 scale used in the CRS is a categorical measure?

A: You do not have to, because even if the test increases in difficulty, the benchmark survey already captures this. Also, keep in mind that although the 0–10 response scale in the CRS is a categorical measure, we are utilizing deviation scores to establish the threshold values. These are ratio-level measures—meaning they can be subjected to all manner of statistical operations.

Q: How do I ensure against social bias and contamination of results?

A: If there is a fear that the subset of users providing the baseline might inadvertently reveal the test, an approach to conceal it would be to develop many pen tests but deploy just one of them. On the other hand, if you believe that users are unlikely to respond honestly, ensure that you compare the CRS data against other behavioral data (such as email opening to clickthroughs) to crosscheck their responses. Besides, not getting honest responses is an indicator that perhaps the users in the organization don't trust the IT department. This can be changed by communicating the value of the CRS, not by awarding punishments and reprimands based on pen tests, and procuring buy-in from users.

Q: As an IT manager, what's my role in the CRS?

A: Your role is that of a physician who is working to understand your users and come up with solutions that work for them. Your role is also that of the change agent, who comes up with solutions that work for the end user and helps them succeed in their work through effectively implementing these solutions. The CRS allows you to do all this. It provides the toolkit for procuring feedback from users, the framework for diagnosis, the quantitative data for fostering accountability, and ultimately the source of the solutions to help make users cybersafe. It also allows you to devise solutions and track their success.

Q: Does a good NRS mean I have achieved cyber resilience?

A: Not completely. To the extent that phishing is the only threat to your organization's security, you have achieved resilience over it, but cyber-attacks target other user weaknesses. There are many more vulnerabilities that users' actions can create, any of which can be co-opted by social engineers. To thwart those, users must possess good cyber hygiene. But what does good cyber hygiene even entail? And how do organizations assess what users need and already possess? This is discussed in chapter 8.

# 8 FROM CYBER RISK TO CYBER HYGIENE

No conversation about users' cyber risk is complete without a discussion of cyber hygiene. Improving it is the justification behind many user-focused initiatives. From awareness training to pen testing, patch management, VPN use, strong password use, two-factor authentication, and antivirus use, all are promoted as necessary because they improve users' cyber hygiene.

But what is cyber hygiene? What are its components? How should we measure it? Why should it even be measured? And how does cyber hygiene fit in with the CRS-based assessment? The answers to all these questions are presented in this chapter.

## WHAT IS CYBER HYGIENE?

As a concept, cyber hygiene is simple to understand. It logically weds the actions people could take to protect their data and devices from compromise to the personal hygiene actions people routinely enact to safeguard and improve their health. There is a certain catchiness and obviousness to the term *cyber hygiene*. This has driven its widespread adoption, with an online search for it returning over 30 million pages. The call for cyber hygiene appears in news articles, policy manuals, congressional testimonies, advocacy sites, and even military doctrines.

But while there are millions of pages that use the term, until recently there was little clarity on what it really meant. There wasn't much work done defining it or clarifying the actions that led to it. In a study comparing cyber hygiene practices across European Union member nations, the European Union Agency for Network and Information Security (ENISA) found that while cyber hygiene was considered analogous to personal hygiene, there was no single standard or commonly agreed approach to it.[1]

Thanks to this metaphorical connection to health and hygiene, cyber hygiene was being promoted as efficacious and protective. Just like brushing teeth daily, washing hands regularly, or taking vitamins, that people should do this or that online became the mantra of security advocates. One technology blog, Cybersecurity Is Cyber Health, took it even further.[2] It equated every cyber hygiene action to something in human hygiene. The use of obsolete software was compared to poor heredity, the lack of technical safeguards to not being vaccinated, visiting unreliable websites to being promiscuous, and so on. Online user actions were even linked to psychological health, fetal ultrasounds, newborns, and unwanted pregnancies.

Such metaphorical equivalences were not just wrong; they were making users more vulnerable. (I outlined the reasons in a paper titled "Stop Telling People to Take Those Cyber Hygiene Multivitamins," which is available online.[3]) I briefly present some of the reasons here.

First, at the conceptual level, how we think about a problem influences how we address it. The development of airplanes is a good example. Because the flying capabilities of birds appeared self-evident, since antiquity, our mental models of flying have been based on avian flight. This inspired ancient fables—Greek stories of Daedalus and Icarus mythologizing the use of birdlike wings for human flight—and shaped twentieth-century attempts at fabricating aircraft with wings that flapped and ended up flopping.

A similar problem stymied the development of germ theory. For centuries, most scientists believed that noxious air, or miasma, was the cause of diseases such as malaria and dengue. Bad air could be smelled; it was obvious and unappealing. Disease-causing germs and microorganisms weren't. The miasma view led to open-air sanitoriums, where patients took walks in fresh air, which exposed them to more of the same disease-carrying mosquitoes, reinfecting them and prolonging their illness.

Problem conceptualization is also important because it shapes the types of solutions we develop. Today, the conceptualization of cyber hygiene in human health terms has promoted the development of USB condoms and antivirus-type phishing detection programs. They extend the analogy. Because of this, rather than designing safer USB ports and email protocols, security engineers have focused on what users should do instead.

Another level on which logical analogies hurt is that they foster a certainty or mastery over a problem. They make us believe we understand the issue, so we move on toward solving it. For centuries, inventors believed they understood how things flew, and scientists believed they knew what caused diseases. Following this, aircraft makers focused on thin flapping wings, and convalescence therapy advised sick people to spend more time outdoors. This delayed the progress of science and caused damage to life and limb—because no one stopped to consider whether their knowledge sufficed.

In cybersecurity, leading organizations such as the National Institute of Standards and Technology (NIST) have done likewise. They believed they understood the problem of cyber hygiene, following which they developed best practices and then asked everyone to adopt them. An indicative example is the 2004 NIST directive to implement long and strong passwords with a range of numbers, letters, and special characters that expired every three months.[4] Microsoft and other corporations widely promoted the idea. They redesigned their software so passwords expired every three months, and users were forced to comply.

But the NIST developed the policy by testing how long it took for computers to crack passwords, not how users would revise and remember them across hundreds of different online web services. This led to enormous confusion for users, who were constantly bombarded with password expiry notifications that they had to comply with in order to regain access to each service.

To keep up, many users reused passwords, while others blindly complied. This became the perfect hook for phishing attacks, which began mimicking password expiry notifications. Recognizing this, in 2017 the NIST finally reversed the guideline, but this was a decade after the damage had already been done. Users had been routinized into believing password expiry warnings, reusing passwords, and clicking on links in such emails. To this day, many organizations worldwide continue to insist that users change their password every few months, because of which phishers persist with this hook and will likely continue to do for years to come.

The third problem that comes from thinking of cyber hygiene in human hygiene terms is that practices such as washing hands, unless done at an extreme and obsessive level, have no downside. But the same isn't true for cyber hygiene. Blindly applying a practice suggestion because some security experts say it is protective can enhance risk. Let's take the example of the secure socket layer (SSL) icon (a lock) that many security training programs instruct users to look for next to a website's name in the address bar of their browser.

The first problem is that users don't really understand what the SSL certification on a website means. Many think it means the website is authentic when actually all it means is that the communication is encrypted.[5]

The second problem with this guideline is that many phishing websites—two out of three, according to a 2019 Anti-Phishing Working Group (APWG) report—also have SSL certificates.[6] Thus, users who apply the SSL best practice get more easily victimized. The same problem plagues users who place too much credence in calls for virus protection use, VPN use, and routine patch and firmware application. The security efficacy of any of these practices is limited. Virus protection and VPNs cannot stop every attack, but users don't realize this, so many engage in riskier actions because they believe in their protections. Adding to this, each of these services can be co-opted by hackers. Social engineering attacks have used every ruse from fake software update notices to antivirus patch notifications, but because users believe in their protections, they throw caution to the wind and comply, which makes them more vulnerable.

Another fundamental difference that makes health and cyber hygiene profoundly noncomparable is what they protect. Personal hygiene protects the human body, but the human body is already resilient. Even without many modern hygiene practices, such as washing hands with soap, humans can ward off many diseases. There have been many news stories and research studies that point to poor hand-washing practices among restaurant workers in the US, yet, except for small outbreaks, their net effect on the spread of disease across the nation has been limited.

This is because, over the millennia, humans have evolved defenses: sensory organs with follicles, hair, nails, eyelashes, cilia, and mucous membranes that trap most intrusions, as well as internal organs with complex immune responses. They function independently, without our control and also in tandem. We develop immunity throughout life, sometimes after infections. We are further protected by the brain's intelligent control (such as when someone impulsively swats a stinging bug or washes their hands because they feel dirty). Because of these protections, most of us can live relatively disease-free lives.

In contrast, while technology is collectively capable of highly sophisticated computational tasks, its core components are dumb circuits. They are not only built without any effective protection, but many of them are also flawed at their core. Take computer processing chips and memory cells, the computer's internal organs. The 2019 identification of Meltdown and Spectre vulnerabilities demonstrated that nearly every computer chip manufactured in the preceding two decades had flaws in its core algorithms, rendering them vulnerable to various exploits.[7] Similarly, dynamic memory cells (D-RAMs) are also vulnerable to leaking their electrical charges as they interact—called the rowhammer effect[8]—which can be exploited in a D-RAM attack to get root access to systems.

The same is the case for the "sensory organs" of computing devices: touch pads, microphones, cameras, and input devices. Each is easily corruptible using key loggers and other programs. Layered on these are many apps, all using different schemes and privileges, which interface with the system's internal organs. Some of these apps are programmed poorly, while others can be manipulated by rogue programmers using malware that can infect everything from the sensory organs of the computer to its internals.

None of the internal or external "organs" of technology can learn to protect themselves, which means an attack on one layer can be repeated until it is discovered and a fix developed. This takes time. Recall the statistics from chapter 1, where I discussed how social engineering evolved: the average hacker spends about six months on a network before being discovered, and 50 percent of organizations targeted by social engineering hacks were attacked again within the year. These are possible because, unlike the human body, technology cannot develop immunity or resistance. A hacker can attack an organization without being discovered and then attack it again after being discovered.[9]

A final reason why cyber hygiene shouldn't be equated with personal health hygiene practices is that errors in security accumulate differently. Because of the lack of evolved protections in computing, a single phishing email with a malware payload can trick users, circumvent many endpoint security protections, and enter the core of a system and gain a foothold, without needing to evolve at each stage. In contrast, take the constantly evolving influenza virus, which kills over 600,000 people and hospitalizes millions more globally each year. Not everyone who is infected from it dies. Some people are resistant to certain strains of the virus, and others have different levels of immunity and inherent protections.[10] This is why, in the presence of someone with the flu, a person with poor hand-washing hygiene still may not get infected.

But let's consider poor hand-washing hygiene as a case in point. What would be the impact on deaths from the flu of a 10 percent failure in hand-washing hygiene—where, say, Americans don't wash their hands for the CDC's recommended 20 seconds? Actually, it wouldn't be much. The reason is that 97 percent of Americans already don't wash their hands enough.[11] It clearly hasn't affected the death rates from the flu, which, until COVID protocols reduced transmission and the rate of infection even further, had remained more or less the same for more than a decade.

In contrast, a 10 percent failure rate for SSL certificates could have an entirely different outcome. If these certificates are used in email-based phishing attacks with a 10 percent relevance rate (users for whom the content is relevant), on an email network that allows 10 percent of these emails through to the end user, with just 10 percent of the users clicking and enabling the malware, the probability of each error would magnify geometrically. It becomes a dependent probability that can be computed by the formula $(1-.90^{\wedge k})$, where $k$ is the number of layers of vulnerability. The probability of a breach taking place would be 34 percent. These are conservative estimates. As I noted earlier in the book, actually 30–70 percent of phishing emails are usually opened, and there are many rogue SSL certificates and fake web pages on the internet. Thus, unlike in the human body, each potential failure because of a flawed hygiene practice significantly magnifies the overall risk in computing.

It is clear from these arguments that personal hygiene and cyber hygiene are not at all analogous. We just cannot afford the same leeway with cyber hygiene. We need greater precision in how we define it and identify what it means and what it doesn't.

Because of its conceptual similarity to personal hygiene, however, no one really stopped to explain it or think it through. Instead, we have had blanket policy suggestions that were never really thought through. Everyone believed they understood the concept of cyber hygiene, so no one developed a valid measure for it. Because of this, no one knew who had cyber hygiene, who lacked it, or even what they lacked.

Since cyber hygiene was a buzzword used to castigate and blame users, usually after a breach, none of this mattered. It was part of the collective consciousness. You could always point to someone online not doing something correctly or enough and say they lacked cyber hygiene. That is why I am forced to use the term here even though I am critical of it.

But I wasn't willing to overlook cyber hygiene's flaws. I chose to correct them. I committed part of my research program to improving the conceptualization and measurement of cyber hygiene. This began at the foundations of the concept.

## THE ROOTS OF CYBER HYGIENE

To get to its roots, I interviewed CISOs, security advocates, technologists, and IT staff at various organizations. I asked each to list out specific user activities that they believed exemplified cyber hygiene and how they helped the organization's security. Their answers varied. For some, good practices included specific actions such as applying patches and updating antivirus. For others, cyber hygiene entailed users' thoughts such as checking the SSL encryption icon or checking to ensure a webpage was authentic. For some others, it involved training, awareness building, pen-testing, and promoting cyber safety. And for still others, it encompassed everything users did or should know to do to protect the enterprise from cyberattacks. But while the activities they listed differed, they all agreed on one thing: Cyber hygiene activities helped build cyber resilience.

Cyber resilience—making the organization impervious to cyberattacks–was the singular driver. It was the desired outcome from every cyber hygiene practice, action, thought, and process. It was at the root. This meant that any definition and measurement of cyber hygiene had to be aligned with this ultimate purpose. It had to help in building cyber resilience in the organization. This meant changing users' poor thought patterns with hygienic alternatives. It required managing change.

To understand how best to do this, I turned to research on Organizational Design—a field of management science developed in the 1940s by Kurt Levin at MIT. The field has developed well-known organizational change solutions such as Total Quality Management and Six Sigma, which have been applied by leading organizations the world over to manage change,[12] but has never been extended to cyber security planning. Based on the literature, I identified five principles for building cyber resilience, each representing a layer of minimum requirements that a cyber hygiene activity must meet (figure 8.1).

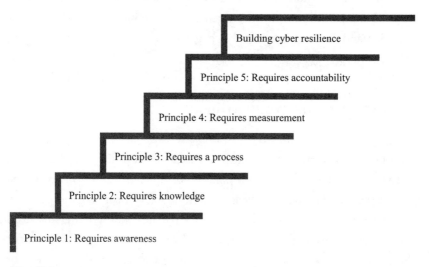

**Figure 8.1**
Five principles for building cyber resilience.

### Principle 1: Building Resilience Requires More than Creating Awareness

I was recently at a conference in the UK debating the limits of security awareness training. Someone in the audience asked me why, when his mother now knows about the Nigerian phishing attack, I would claim awareness training doesn't work. His contention was that this was proof enough for him that she was resilient to such attacks, a common mistake people make. Awareness, while necessary, is far from sufficient. It involves familiarity with a concept, which is not the same as knowledge or understanding.

You can be aware of something, such as a tornado, and have no clue what it means or what to do when you are in its path. Likewise, you can be aware of a Nigerian phishing email but have no understanding of its potential capabilities. This is important because not all Nigerian phishing emails are pretexting attacks. Many have evolved to port malware, appearing as Eastern European phishing emails, Microsoft help-desk emails, or phone-based attacks. Some are used to collect information, which means even opening the email signals that the email account is active and in use. Knowledge is essential for dealing with each of these threats appropriately. Opening, answering, clicking, deleting, or opening on another device will each define whether the outcome is benign or harmful. They provide hackers with insights into the user's behaviors that can lead to even more focused attacks. Thus, awareness is necessary for building cyber resilience, but it's the minimum requirement.

### Principle 2: Building Resilience Means More than Imparting Knowledge

Whenever awareness doesn't suffice, the natural next thought in most people's minds is that users need knowledge. However, knowledge is a broad concept. It ranges from know-how to in-depth expertise. For cybersecurity, the knowledge required can range even further, based on the technology being used. As noted in chapter 4's discussion on cyber risk beliefs, computing technology has far too many layers, and it's impossible for any person to master all of them. Instead, achieving cyber hygiene requires knowledge that is just adequate for the task. It requires that users be *sufficiently* knowledgeable to achieve a *satisfactory* level of resilience. That is, users need knowledge that *satisfices*—a term from cognitive science that combines the preceding two terms.[13]

This is distinct from some present approaches to cyber hygiene, where awareness training programs teach users to achieve mastery by examining dozens (more than 40 in one training program) of different features in emails to decide whether the email is deceptive. This is impractical for users who juggle multiple emails, email accounts, devices, and competing cognitive tasks. Such knowledge-intensive approaches do little more than subject users to a standard they cannot possibly achieve. Over time, this reduces their motivation to even try to be safe online, which reduces resilience.

Instead, building cyber resilience requires an examination of what satisfices for each decision. This has to be done at the user level, by examining what users are doing and matching their task requirements with security knowledge. In the case of email security, users don't need to understand IP addresses, malware signatures, ports, macros, and transmission protocols. They need to know where to look for the major signals in the email that should trigger suspicion and know the steps to take after that. This satisfices.

### Principle 3: Building Resilience Is a Process

The third principle moves beyond awareness and knowledge to its constituents. There isn't one thought or action that IT departments can instill that will lead to resilience. Resilience is the outcome of a series of safety approaches, each complementing the others, the use of which is encouraged, whose optimality is frequently ascertained, and whose capabilities are constantly updated to meet the ever-evolving changes in the threat landscape.

Moreover, resilience is not a binary outcome. It is a process that occurs over time, as users change how they act and think gradually and in degrees. This requires creating a culture of cybersecurity, one that involves more than just dictating best practices. For users, the culture of cybersecurity must be something they are vested in, find valuable, and see as part of their work process.

### Principle 4: Building Resilience Requires Measurement

If we ever hope to create a fundamental change in how users think and act on their devices, we need to ensure that there is an equitable process of building resilience. A core necessity is measurement—the ability to quantify, track, and show what is lacking and what's already been achieved. The measurement cannot be convenient. That is, we cannot simply use what we have at hand and assume that it is all we need. This is the case with most awareness training data, which uses pen-test results. Not only is this data limited, but it also isn't comprehensive enough to account for all the satisficing actions of users.

The measurement used to build cyber resilience must be comprehensive. It must account for the different aspects of cyber hygiene that directly or indirectly contribute to resilience. To protect credentials, for instance, users need to be aware of the need for complex passwords. They must also know how to create and store them and have actually done so. These are three different facets—awareness, knowledge, and behavior. On top of this, users may need to have the ability—the capacity—to store these passwords safely. In order to build cyber resilience, we must cultivate all four of these facets across a variety of user functions (email, messaging, browsing, and so on). The measurement must also be capable of capturing their intensity in all its gradations (i.e., how much awareness, knowledge, etc.). This way, we can track users, identify their shortcomings, compare them, and improve them. Thus, comprehensive measurement is necessary for building cyber resilience.

### Principle 5: Building Resilience Requires Accountability

Developing a culture of cybersafety requires that users buy into the process. This cannot be achieved from the top down. All that does is instill fear or foster compliance without thinking. It fosters a rigidly compliant user population that would rather avoid using company emails for fear of being caught in a phishing test. It reduces employee efficiency and doesn't lead to a dynamic or agile workforce.

What we need instead is accountability that is data driven, transparent, and provides the user with actionable metrics for improving how they think and act online. This requires measurement that can be communicated to users and understood by them. When sound and objective metrics are applied consistently and fairly and are shared with users in the organization, accountability takes hold. This ultimately helps achieve cyber resilience.

These five principles lead to the conclusion that achieving cyber resilience is beyond just awareness and knowledge; it requires measurement. What is necessary is a measure that is broad in coverage, considers more than just an action and reaction (as in a phishing email and its response), and covers the entire gamut of satisficing actions that users can perform to prevent cyber breaches. In 2020, I developed a measure that meets all these objectives, which I called the Cyber Hygiene Inventory (CHI).

### MEASURING CYBER HYGIENE

To develop the measure, I used a modified version of the concept mapping technique developed by Cornell University professor William Trochim. People who hear the term *concept mapping* often confuse it with brainstorming, but this is different. Concept mapping is a quantitative process involving multiple steps that culminates in a shared identification of a construct, its dimensions, and its constituent measures. It uses a combination of scale-development processes (nonmetric multidimensional scaling, cluster analysis) and qualitative approaches (allowing user involvement in defining the concept boundaries and dimensions). The full details of the process are in my peer-reviewed paper referenced in the notes to this chapter.[14]

Briefly, the process involved multiple stages, where the concept of cyber hygiene was first defined independently by different security experts. Next, individual questions that reflect the construct were developed. There were hundreds of potential questions, which were then sorted into various groups by IT professionals and subjected to different levels of statistical analysis. This process revealed the core dimensions of cyber hygiene and the questions that defined each dimension. Individual questions were then tested, refined, validated, and finalized using a series of surveys on different user populations. The outcome of the process was a consensus definition on what cyber hygiene entailed, its various facets or dimensions, and a 25-question Cyber Hygiene Inventory.

The definition of cyber hygiene that emerged was that cyber hygiene involved "the cyber security practices that online users should engage in to protect the safety and integrity of their personal information on their Internet enabled devices from being compromised in a cyber-attack."[15]

At the operational or measurement end, there were four levels at which this needed to occur: awareness, knowledge, technical capacity, and enactment. First, cyber hygiene required *awareness*. This involved familiarity with threats and related practices. It required *knowledge*, as in satisfactory understanding. These were cognitive or mind-level factors. They needed to be complemented with technical *capacity*, or the requisite tools. The final level was the *enactment*, the actual performance of the action itself.

Each level of cyber hygiene built on the others, with awareness being the bare minimum and enactment being the most desired. They also occurred in degrees, so you could possess each of them to a greater or lesser extent. Altogether, they addressed the principles necessary for building cyber resilience.

Twenty-five cybersafety-related thoughts and actions were identified as being relevant across user populations. These fell within five broad areas of cyber actions and were organized using the acronym SAFETY: *S* pertains to storage and device hygiene, *A* signifies authentication and credential hygiene, *F* signifies friend connections and social media hygiene, *E* pertains to email and messaging, *T* stands for transmission hygiene, and *Y* stands for "you" or the "self," signifying the user's responsibility in ensuring cyber hygiene.

Altogether, the 25 best-practice cybersafety thoughts and actions measured across the four levels are called the Cyber Hygiene Inventory. Each cyber hygiene practice in the CHI is a survey question. Users answer each question using a five-item, 0–4 response scale. This allows the cumulative CHI scores to range from 0 to 100, with higher scores being more desirable. From the 25 questions, 5 questions each measure storage and device hygiene, transmission hygiene, friend connections and social media hygiene, authentication and credential hygiene, and email and messaging hygiene. The CHI is presented in table 8.1.

The CHI is like no other measure. For one thing, there was previously no valid measure of user cyber hygiene. It is the first that conceptualizes user cyber hygiene and identifies its facets and dimensions.

Second, the CHI can be used across different organizations and user populations, no matter the size of the organization or the complexity of the users' IT function. It provides a baseline of the minimum necessary hygiene levels and can be easily scaled up, in that more questions can be added to it, to measure the more complex needs of specific user groups.

**Table 8.1**

The Cyber Hygiene Inventory

| Question | Cyber Hygiene Inventory (CHI) | What each question measures |
|---|---|---|
| 1 | how to check your device to ensure it has the latest operating system, software update, or patch | storage and device hygiene |
| 2 | how to remove sensitive data being stored on your device | storage and device hygiene |
| 3 | how to run a virus scan on any new USB drive or external storage device | storage and device hygiene |
| 4 | how to keep virus protection updated | storage and device hygiene |
| 5 | how to enable firewalls on your computing devices | storage and device hygiene |
| 6 | how to enable two-factor or multi-factor authentication for logins | authentication and credential hygiene |
| 7 | how to change default passwords on all internet-enabled devices | authentication and credential hygiene |
| 8 | how to change the default username from "administrator" to something unique on all internet-enabled devices | authentication and credential hygiene |
| 9 | how to manage how your browser stores passwords | authentication and credential hygiene |
| 10 | how to store your logins and passwords on a password management app | authentication and credential hygiene |
| 11 | how to reassess who you are connected to on social media from time to time | friend connections and social media hygiene |
| 12 | how to know who you are connected to on social media | friend connections and social media hygiene |
| 13 | how to assess the authenticity of social media friend/information requests | friend connections and social media hygiene |
| 14 | how to curate your social media profile information about yourself | friend connections and social media hygiene |
| 15 | how to manage access to sensitive personal information (such as your location) when posting or sharing on social media | friend connections and social media hygiene |
| 16 | how to check an incoming email's header | email and messaging hygiene |
| 17 | how to check whether email requests have grammatical or typographical errors | email and messaging hygiene |
| 18 | how to check a sender's email domain name | email and messaging hygiene |

| Question | Cyber Hygiene Inventory (CHI) | What each question measures |
|---|---|---|
| 19 | how to validate or check the authenticity of a hyperlink or attachment you receive via email | email and messaging hygiene |
| 20 | where to report a questionable email or SMS message | email and messaging hygiene |
| 21 | how to manage computer discovery on public networks | transmission hygiene |
| 22 | how to restrict who can connect to your device via Bluetooth | transmission hygiene |
| 23 | how to place online alerts for your name or personal information (e.g., Google Alerts) | transmission hygiene |
| 24 | how to check the quality of the SSL certificate on a website before entering sensitive personal information | transmission hygiene |
| 25 | how to connect to the internet using a VPN | transmission hygiene |

*Note*: Each question indirectly translates to a best-practice suggestion. When applying the CHI, assess four facets of user cognition and behavior: awareness, knowledge, technical capacity, and utilization. Individual questions should be measured using a 0–4 response scale. We can assess them at the subjective level (as in perceived knowledge, perceptions of technical capacity/availability, and intent to use). Alternatively, knowledge, capacity, and utilization can be measured at the objective level with true/false questions; capacity and utilization can also be observationally coded.

The overall 25-item scale when cumulated using a 0–4 scale nets a range of responses from 0 to 100. An earlier version of the CHI consisted of 20 items and used a 0–5 response scale. Five questions were added to improve coverage and balance the scale. While the final 25 questions should suffice in most cases, questions might need to be reworded or added to suit the organizational context. Although doing so may improve ecological validity, it could alter the overall outcome.

Third, the CHI is technology and product agnostic. It doesn't measure digital safety for a specific type of browser, app, operating environment, or software. Instead, it focuses on users' cognitive and behavioral readiness. Thus, it can be applied anywhere in the world and over time to measure baseline cyber hygiene without needing much modification.

Fourth, the CHI is measured using standard survey questions that most people in the world know how to answer. Interpreting the survey responses is straightforward and easy for both the IT staff and respondents.

Fifth, the CHI is transparent in what it measures and how it arrives at its score. This is in contrast to approaches such as the Department of Homeland Security's cyber hygiene scoring dashboard. Their score is based on an undisclosed number of factors, which they purposefully conceal to protect them from falling into the hands of hackers. This, however, makes it hard for the end user, be it the organization's IT department or the employees, to be held accountable for them, because the user never really knows what they did right or wrong that netted them the score. In contrast, the CHI makes it easy for IT and users to know where the gaps exist; they know where corrective actions are necessary and why they are being held accountable.

Finally, the CHI uses a range of question types. The approach isn't limited to a type of question or a single factor of user safety or its performance. It captures knowledge, experience, and the technical factors of hygiene. It utilizes subjective questions for the cognitive and behavioral measurement and uses objective questions for measuring the technical capacity available. By capturing facets of users in more ways, it provides a more comprehensive assessment of cyber hygiene.

Now that we know the origins of the term *cyber hygiene* and its advantages, let's examine how cyber risk and cyber hygiene work together to achieve cyber resilience.

## THE RELATIONSHIP BETWEEN CYBER HYGIENE AND CYBER RISK

The Cyber Risk Survey measures risk from phishing and other forms of social engineering. Such attacks are capable of inflicting damage directly by making incursions into devices and systems. Phishing is the most common and successful approach used to do so.

Compared to this, the Cyber Hygiene Inventory captures a broader range of behaviors that protect users. It includes social engineering as well as other types of attacks. For instance, transmission hygiene tracks whether users know how to manage device discovery on networks. This does not directly guard against email-based phishing but could protect against other social engineering vectors, such as man-in-the-middle (where a hacker eavesdrops on communications) and Wi-Fi spoofing (where the hacker creates a fake access point). Likewise, keeping virus protections updated, a form of storage and device hygiene, does little to protect against phishing but helps ward off other forms of device and network disruptions. Thus, the CHI captures a broader swath of thoughts and actions that lead to cybersafety.

Both the CRS and CHI also differ in how they measure users. Given its focus, the CRS is a reactive measure. Some of what it measures, such as cyber risk beliefs and suspicion, remains stable over the short term, but other measures, such as heuristic processing, are entirely contingent on the situational factors. In contrast, the CHI is a stable measure of the facets of cybersafety users are aware of, are competent in, and have the capacity to deal with.

Both the CRS and CHI are, however, complementary measures and together tell the complete story. If we think of phishing as a cough or sneeze that carries the flu virus, the CRS captures the individual's propensity to contract the flu. The CHI captures the protective actions that people can take to avoid getting infected or spreading the flu virus. Its protections aren't limited to just the flu. It can help ward off many more disease-causing germs as well.

Just as in human health, hygiene alone doesn't guarantee against infection. It simply reduces the chances of easy infection. There are always cases where it doesn't prevent infection. For instance, there are people with excellent personal hygiene who may get infected by the flu because of a compromised immune system or because the disease appeared through an unexpected source (such as through contaminated food). At the other extreme, there are also people who despite poor hygiene might avoid the flu because they haven't been exposed to it or have a strong immune system that makes them resistant.

The same is true with cyber hygiene and cyber risk. While most people with good cyber hygiene should have lower cyber risk, there will always be users with good cyber hygiene who might still fall victim to phishing because of nonconscious behaviors that aren't measured by the CHI. On the other end, there could also be users with low cyber hygiene who are protected from phishing. This would be the case if an organization's IT department creates email delivery rules (e.g., do not deliver any email marked External), which users aren't aware of, that protect users against phishing but aren't accounted for in the CRS.

But while the nonconscious behavior of users might be missed by the CHI, it would be accounted for by the CRS. Likewise, if the CRS misses a company-wide email delivery rule, the CHI could capture it in the capacity dimension. In this way, the two measurement approaches complement each other—together capturing the full range of causes. Such lessons are hard to acquire without measuring cyber hygiene and cyber risk. Thus, measuring both the CRS and CHI leads to a comprehensive estimate of cyber resilience.

An IT manager wishing to build user reliance should therefore use both measures. Rather than simply focusing on phish testing (as most organizations currently do) or even just generating a user risk profile, IT management can learn more by applying the CRS and the CHI. The former needs to be conducted more often to accommodate evolving attacks. The latter needs to be implemented at least once each year, or twice in circumstances where it is used for tracking improvements. Depending on the goal of the organization, there are a number of ways the CRS and CHI can be implemented together. The following use cases demonstrate the possibilities.

**Use case 1: Using the risk index to assess individual gaps and the hygiene inventory to understand overall lack**    This is the most basic use of the CHI and CRS. It involves implementing the hygiene inventory and then conducting a risk assessment using the approach discussed in chapter 7. If the cyber hygiene measurement is conducted prior to risk assessment, it helps determine the baseline levels of user awareness, knowledge, capacity, and enactment of SAFETY practices.

Subsequent risk assessment using the CRS can then be used to perform a gap analysis that can help pinpoint gaps in hygiene that led to risk. For instance, the analysis can compare awareness and enactment across the five SAFETY dimensions of cyber hygiene for individuals in the high- and low-risk groups identified using the CRS.

This could help identify whether awareness or actual use of different cyber hygiene practices contributed to their cyber risk. The combined use of the hygiene measures and the CRS can also help in understanding how big the risk from phishing is to overall resilience. For instance, knowing that two-factor authentication (2FA) wasn't being appropriately used combined with the overall risk profile of users could help in understanding how much of a threat phishing attacks that targeted credentials posed to the organization.

**Use case 2: Using the CRS and CHI to build a culture of cyber-safety**  The CRS provides IT departments with a tool for measurement and quantification. This, when communicated appropriately, can help usher in a culture of individual responsibility for cyber incident protection.

Using the CRS's open-ended responses, IT staff can also identify good practices—actions that keep people at a lower risk level from getting phished—and create best-practice rituals for others in the organization. The efficacy of such practices in fostering new cybersafety thoughts and actions can be evaluated using the CHI. The cyber hygiene scores along with the risk score can also serve as a starting point for discussion—necessary steps for building a culture of cybersafety.

The scores can be used to communicate with users. Informing users about their cyber hygiene scores along with their risk profiles can help them understand where they are lacking, what they need to think about or do differently, and how their actions detract from or contribute to securing the organization. The scores can thus provide a baseline for discussion with users and for creating a road map for where they need to go.

For IT managers, the two indexes can provide a variety of metrics. Using the hygiene score of users, IT managers can set a target for awareness or knowledge and then provide the necessary tools for users to achieve it. Next, by using the CRS, they can determine whether the risk goals were met, whether awareness and knowledge contributed to achieving them, and how they did. This can help fine-tune safety initiatives and make them more effective. Finally, using the CHI, IT managers can determine who should or shouldn't have access to various files or servers. Access privileges can be extended or revoked based on individual performance on the CRS, and users can be given a road map for improving their scores in order to get access.

In this way, the CRS and CHI can serve as tools for communication, discussion, providing feedback, and tracking. They can help engage users, involve them in cybersafety improvement, and usher in a sense of shared responsibility toward protecting the enterprise from cyberattacks.

**Use case 3: Using the CRS and CHI for strategic planning—deciding who gets what, when, and why** There exist dozens of different cyber-safety initiatives, services, and solutions, and each has a strong case in favor of its implementation. The question for IT managers is how best to use them. Which of them work, and how much should the organization spend on each? Which ones are relevant? Who in the organization needs them most? Why? These are questions of strategy that can avoid needless expenditures and save time and effort.

The CRS along with the CHI can help IT departments understand who needs what solution based on their overall risk estimations. The impact of each solution on the different dimensions of cyber hygiene levels can then be used to perform a cost-benefit assessment for each initiative. Some initiatives may even have a permeating effect across different dimensions of hygiene so while they cost more, their impact would be wider. These initiatives can now be identified.

Furthermore, while organizations routinely conduct awareness training programs, who needs more awareness and what subject matter the training should focus on remain unclear. Many IT departments follow a vendor's lead and craft programs based on their advice, but such guidance might not be suited for every organization. Now, using the hygiene inventory, IT departments can measure what is most necessary and focus on just the SAFETY dimensions that matter. They can even focus on individual factors within them. For instance, knowing that users' knowledge of messaging hygiene is low and that this contributes to high cyber risk can lead to the development of initiatives that focus on just this issue. Alternatively, knowing that there is a lack of enactment might help in developing initiatives that foster more use of safety technology, which might not require investing in more awareness training.

Another question every strategic planner contends with is the relative value of training, both in terms of time and cost, versus solutions that help contain and constrain (discussed in chapter 3). Today, there is no measurable mechanism for testing the relative efficacy of various alternatives. All that IT departments have is the data supplied by vendors or from their own experience contending with an attack, where the value of individual systems becomes clear.

Thanks to the CHI, all the organization needs to do is conduct a cyber hygiene assessment once a year and follow up once more after conducting phishing tests using the CRS. Assuming that the pen test was the only factor influencing the two applications of the hygiene inventory, any changes in the CHI could be attributed to it.

In a similar vein, rather than resorting to blaming users for all problems, now IT departments can find exactly where the problems are by contrasting the risk among the users with the capacity measure of the hygiene inventory. This can help pinpoint areas where users do not have the resources for being cybersafe. Once determined, budgets, program resources, and safety initiatives can be developed to improve this area. Their success can also be tracked by simply repeating the entire process later in the year. Doing so can save significant time, cost, and effort in protecting users and making the organization cyber resilient.

**Use case 4: Using the CRS and CHI to track how users are improving over time**    For IT managers, knowing whether their users' security posture is improving is pivotal. Awareness initiatives often fall short because they are too focused on building awareness rather than knowledge; at other times, knowledge and awareness might appear high, but cyber risk may continue to rise. As of now, there is no mechanism for pinpointing any of this.

The cyber risk assessment process provides an alternative. By using the CRS, IT managers can understand who is at risk. The CHI goes one step further and allows IT managers to know where the user is lacking and where they are improving over time. Using the CHI, levels of both knowledge and awareness can be tracked along with risk, giving insights into who is at risk, by how much, why, and what else is causing this risk.

Adding to this, IT departments can examine areas where capacity is lacking and provide it or where user actions are required and foster them. For instance, if users are knowledgeable about 2FA but are not using it, then training to improve awareness and knowledge would be a wasted effort. IT departments can instead use these resources to build incentives such as, say, offering a small prize to the first few employees who activate and use 2FA or who use it each time while logging into their email systems. In this way, IT managers can track users across multiple dimensions of risk and behavior and use an evidence-based process to improve their cyber posture.

**Use case 5: Using the CRS and CHI to improve phishing pen-test development**  As discussed in chapter 6, many phishing tests are developed ad hoc. With the risk assessment approach and the V-Triad, IT managers can now design a valid test. But what must this test measure? Most pen tests either copy templates provided by vendors or use a few broad themes (e.g., retail shopping email spoofs, generic business email compromises). But what facet of user vulnerability should tests target?

Now, with the CHI, tests can target specific areas within the five SAFETY dimensions. They can assess the extent to which users' risk is influenced by them and develop focused pen tests that examine the specific levels (as in awareness, knowledge, and such) of user readiness. So, if transmission hygiene is low, IT departments can craft pen tests that focus on Wi-Fi spoofing and other facets of transmission. If social media hygiene is lacking, we can create attacks that test users' readiness on that dimension. In this way, using the CHI to guide test development and using the CRS as a measuring stick, IT departments can build tests that are meaningful. They can also track whether after the test risk went down and whether hygiene levels in the specific factors that the test or training addressed improved.

**Use case 6: Using the NRS and CHI to compare different groups of users within a division or within the same organization**  Much of the preceding discussion focused on risk assessment of a single group of users who vary in their cyber risk. IT managers can use the same process across different divisions or use additional markers to compare users within the same division to identify functional areas, organizational divisions, and demographic variances in readiness and the reasons for them.

This helps test many of the biases that exist. For instance, we often hear of IT staff castigating certain groups of users, such as those who may be less technologically savvy because of their age or tenure in the organization or those working in certain functions, such as clerical staff, for being careless when online. Now, using the CRS, IT departments can derive the net resilience score (NRS) for that group of users and, using the responses to the CRS, identify the reasons for the risk. Then, using the CHI, IT managers can cater solutions to the division or group's need and calibrate and track their improvement over time.

**Use case 7: Using the two measurement approaches for assessing insurability in a sector** Companies providing cybersecurity insurance need to understand the security postures and resilience levels of organizations. This is pivotal for actuarial computations and for setting insurance premiums. Presently, this is done using audits of technology and the examination of training and pen-testing data provided by individual organizations.

While the risk assessment process previously discussed that leads up to the NRS requires participation by each organization's IT department, the hygiene inventory needs just a survey and can be administered directly by insurance companies. This provides two sets of rich data: risk data and safety data. Together, the two data streams can reveal the state of preparedness in a sector and where different organizations rank on it.

As an alternative, cybersecurity insurers can also develop their own phishing pen-test exemplar using the V-Triad and assess its baseline using their in-house "expert" sample or by aggregating the results of baseline tests conducted by some of their current customers. New organizations looking for insurance can then be asked to conduct a pen test using the exemplar, and their data can be compared to the insurer's internal baseline. This way, the security insurer can independently derive an NRS and get a more accurate estimate of the level of resilience of the organization. This data can then be contrasted with the organization's CHI results to help the organization identify areas of improvement that could reduce their insurance premiums.

**Use case 8: Using the two approaches for comparing different sectors, regions, and nations** Finally, law enforcement and public policy makers can apply the hygiene inventory across a sample of organizations to understand what is lacking in a sector or even across a region or nation. A sample of participating organizations can volunteer deidentified internal data that can reveal risk and hygiene levels in the region. This data can serve as a benchmark for deciding how much hygiene is desirable. Using web or telephone surveys, representative data on hygiene levels can then be collected from a sample of respondents in a region and contrasted against the benchmark. This can help build resilience against cyberattacks. The data can inform policy and serve as a tool for allocating law enforcement resources, informing public opinion campaigns, and developing tools and reporting portals in the region.

The eight use cases demonstrate what's possible. They show how different groups—IT managers, policy makers, insurers, and security training organizations—can implement the CHI and the CRS to arrive at actionable insights. The CRS can be used to develop a risk profile of users and explain the reasons for their risk. The CHI can also be used by itself to ascertain hygiene levels of users. But, by using them together, IT departments can garner more granular information about user vulnerability and readiness. Cyber resilience comes from assessment, quantification, and implementation of smart, evidence-based solutions. Using the CRS and CHI, IT departments can stratify users based on their cyber risk, identify the weak links, understand the reasons for their risk, and implement, track, and improve their cyber hygiene solutions. In other words, they can achieve cyber resilience.

Chapter 9 provides examples of organizations that, following the approaches in this book, achieved such resilience.

# 9 A TALE OF FIVE IMPLEMENTATIONS

This chapter examines the implementation of cyber resilience at five organizations: Masonry Inc., a large regional bank headed by its charismatic chief information security officer (CISO), David Marcus; Ashmore Medical Center, a Connecticut hospital headed by a physician-security leader, Alex Sanchez; the Tariff Advisory Commission (TAC), a midsized government organization with over five hundred employees working in the Washington, D.C., area, headed by CISO Elaine Banks; Amazing Income Inc., a fast-growing, dynamic wealth management company with 20 regional offices, protected by CISO Amy Craft; and Akron House, a small nonprofit in northern Pennsylvania with 45 employees, headed by Jake Traposh.

Each vignette presents a case study of CRS implementation and shows how an organization approached and benefited from it. Some cases are detailed and lengthy, whereas others are purposefully short. The goal is to present a variety of implementation approaches that are detailed for those needing it and easy to grasp for those looking for brevity. Each case is that of an actual organization with identifiers and data altered to protect confidentiality. The cases demonstrate how the CRS and CHI can be applied in organizations large and small to craft solutions that don't always end with even more training. As you will see, some solutions will be derived from users organically and converted into organizationally appropriate best practices, others from social science theories that have seldom been extended in the security realm. After reading the cases, you'll certainly recognize, as have the CISOs featured in the cases, that such insights and solutions, only possible because of the CRS, are better than those provided by any security vendor.

## MINDFULNESS AT MASONRY INC.

Masonry Inc. is a large regional bank having over six hundred branch offices across the US, with a powerful presence in the Midwest and Mid-Atlantic regions. Its business interests span all aspects of banking, with retail banking being its most prominent, where the bulk of its more than10,000 employees work. Many of them have direct contact with customers and handle customer data, making each a potential target for social engineering. Losing data, any data, would not just be a nightmare to deal with but would likely destroy the credibility the bank had diligently earned over a century. Hardening them was not just important but pivotal. Ensuring this fell on the shoulders of its CISO, David Marcus.

Marcus had worked in national intelligence and had earned his stripes running offensive operations during the early days of the internet. He had migrated to the defender's side during the days of the Y2K millennium bug and worked his way up the corporate ladder. Thanks to this varied experience, he understood the weaknesses in technology, which made him, in his own words, "adept at knowing what to defend well." But the one facet that troubled him was the social threats to technology, something that seemed to have no solution because it was caused by users—over whom he had limited control, especially in the private sector.

Since joining Masonry in 2010, Marcus had instituted several safeguards, applied many well-known technical constraints that restricted employee access to servers, and embarked on a large-scale awareness campaign. As part of the campaign, he'd introduced a centralized help desk with an email and phone number along with dedicated IT staff manning the desk. Employees could report to the desk anything IT-related they felt was untoward and receive help. Marcus also introduced frequent pen testing, where users were sent mock phishing emails. While many of these attacks were initially conducted in-house, in the last five years, he had sought the help of ethical security hackers (as in white hatters) and other vendors.

The test results were always mixed. Anywhere from 2 percent to 15 percent of different user groups fell for different attacks each time, and there was no logical pattern, let alone rationale, for why this was occurring.

The reporting rates of users seeking help from the desk after a pen test were also dismal, with only one or two people even caring to report.

The most unsettling part of all this was that Marcus did not know why any of this was taking place. Awareness, measured using his available metric—failure on pen tests—was high but naggingly inconsistent: a small but different set of users fell for each attack.

Nothing bothered David Marcus more than this. He liked consistency and wanted to know for sure why this wasn't happening. More than anything else, he wanted a solution that didn't simply involve more training, threatening, or incentivizing. Instead, he wanted to know what was going on, why, and how he could fix it. He had heard about the risk-assessment approach I'd developed and wanted to see if it would help.

Following the approach outlined in chapters 6 and 7, I used the Vishwas Triad (V-Triad) to develop a baseline attack. With inputs from Marcus's IT team, I developed eight attacks, each varying a specific aspect of the bank's internal communication that fit each of the triad's vertices. Beginning with *compatibility*, we focused on re-creating internal routines—printing and file-sharing systems, checking customer accounts—and also introduced *credibility* symbols, such as the name of the network printer, its logos, the colors of the file-sharing portal, and the fonts used to denote the internal customer account portal login page.

The exemplars were all phony in the sense that no such communication was actually being exchanged by these systems. Even when the fonts or colors were accurate, the individual emails had glaring errors, such as presenting the wrong name for the file-sharing portal, presenting the logo of an unapproved browser, and so on. Emails were also *customized* and were addressed specifically to each user.

Next, using 25 employees from the head office, the benchmark for each test was computed using the three questions in the Cyber Risk Survey (CRS). After excluding responses from respondents who weren't representative of the general users in the organization, the answers provided by the expert sample were aggregated and the eight exemplars were rank ordered on quality from easiest to detect to most difficult or capable of deceiving. Two exemplars were scored in the median, considered by most to be neither too difficult nor too easy to detect. The first of these attacks was used in the base-line pen test. The second was used in a follow-up test in the next quarter. The test ranked next highest in difficulty was used in a final follow-up that was conducted three months after the second pen test. Following the procedures outlined in chapter 7, the CRS was used as a measurement tool after each attack, and the diagnosis was used to develop interventions whose imple-mentation success was assessed in each subsequent test, again using the CRS.

For the sake of simplicity, the first pen test was sent to employees working in 18 different branches in the same time zone. This made it possible to better manage the CRS deployment and track the success of interventions by treating those excluded from the assessment as a holdout or control group. The attack mimicked the internal cloud-based file-sharing portal and asked respondents to download an embedded zip file to examine their network printer use overage.

A total of 1,500 users were targeted. Within a week after the attack, every user who was targeted received a personalized email request from David Marcus with the CRS. He implored users to be honest in their responses in order to improve the organization's cyber readiness. The CRS contained a copy of the pen test and of another email that wasn't used in the test, and asked users whether they recalled receiving any of the emails in a pen test and to identify the one they had received.

A total of 1,289 users (85 percent) responded to the CRS. Anyone who had not opened the pen-test email based on the behavioral tracking data or who couldn't accurately recognize the correct email from the two exemplars in the survey was dropped from the analysis. This netted 1,003 respondents, of which 38 percent fell in the high-risk category determined by the baseline cutoff. The remaining 622 respondents (62 percent) were in the low-risk group. The computed net resilience score (NRS) was 24 (i.e., 62–38).

All the users' answers to the CRS open-ended question were then coded using the numerical categorization scheme discussed in chapter 4. Briefly, this involved counting the number of times a response indicated that the user had used a heuristic rule (if-then rule), in-depth systematic thinking (I did this because . . . ), a risk belief (I believed . . . ), or a habitual, pat-terned reaction (I immediately or reactively, inadvertently . . . ).

The occurrences of the sentiments in each user's statements were counted and averaged for the low- and high-risk groups. Then each individual in the high-risk group was compared to this. On average, those in high-risk groups were using relatively more heuristics and had weaker risk beliefs than individuals in the low-risk group. This overall diagnosis became the basis for instituting changes in the organizational education protocols and developing a heuristic and risk belief improvement program.

The improvement program was as follows. The IT department created a brief online video presentation that high-risk users were asked to watch, followed by a short, five-question exercise that ensured they acquired appropriate heuristics that would be available to them in the future. The presentation focused on the internal systems (printers, file-sharing systems, browser, VPNs, and Wi-Fi protocols) and provided them with simple "if-then" rules that they could apply. This was followed by a small exercise where users had to track down and report specific aspects of the various systems for which they had been provided with mental heuristics.

For instance, users were shown which network printers were available in the organization. They were told how such systems would or would not communicate with the user and how such notifications appear. They were then asked in the exercise to report which brand and model of printer(s) they had connected to their system, which of these were network printers and which were used only on their device, and so on. The users in the exercise were asked to fill in a few details about file sharing, ranging from color to the type of antivirus system that the file-sharing system used. As part of the training, users were presented with the approved browsers in the organization and how their various settings could be adjusted to ensure privacy and avoid allowing malicious operators to hijack their system. Following this, the IT department's whitelisting process and how users were being tracked were explained, along with rules of what IT staff would send as communication. The training again reiterated how users needed to report suspicious online activity and then explained the help-desk setup. To reinforce all this, each of these exercises were conducted separately, after the heuristic training, and users had to send in their responses to the help-desk email account set up for reporting suspicious online activity in the organization, which now also served as their training portal.

Three months after the first pen test, the second pen-test email was sent. This attack warned users that their web browsing history violated IT policy. The email, sent from a phony help-desk email account, asked users to click on the embedded hyperlink to examine their browsing history. Only the 1,500 users targeted in the first pen test were sent this attack. Just as in the first test, a week after the attack, all targeted users were sent the CRS.

Looking at the data from the 1,003 users assessed in the first CRS test, just 11 percent were judged as being high risk. The NRS was now $89-11=78$, much improved from 24 at the last test. Once again, following the diagnostic protocol, the open-ended responses were quantified and contrasted. In examining their responses, many of the individuals having the highest risk suffered from poor habits besides some continuing to have bad heuristics. Thus, a habit intervention was planned that targeted everyone in the high-risk category with poor habits, which was administered along with the heuristic improvement program.

The intervention focused on developing cyber mindfulness. The root of mindfulness is in the esoteric Eastern spiritual traditions, and while many in the social science community have suggested it as a solution to user carelessness, none has developed procedures for its implementation. None has identified which vulnerable groups among users are best treated by it. In my past work on college students, those with poor technology habits reacted favorably to certain mindfulness-focused interventions—and I extended some of these approaches to treat the poor habits of users at Masonry Inc.

As part of the mindfulness intervention, users were asked to perform the following tasks: (1) access their email accounts and report the number of external emails they received and responded to each week on their computer or mobile device; (2) track their browser data and report their weekly browsing history; (3) track their data usage by utilizing their terminal's resource monitor; (4) track any or all system notifications (e.g., operating system and other software update prompts, antivirus definition updates). This information, along with screenshots of any of these updates, had to be submitted using an online portal that was linked to the help-desk page.

Each user was required to perform a different operation on a different day of each week, which was preprogrammed into their shared calendar. The reporting was to be performed once a week (or more if the user desired) for four weeks, which prior research on habits has shown to be the period necessary to establish a new routine. The goal of this intervention was to improve users' monitoring of the system-level routines they perform on their computers and devices and break their automatic reactivity. Besides these specific interventions, all users, including those who were judged as low risk, were subjected to a brief online training protocol, where users were told what to look for in phishing attacks—something that had been routinely used in prior training—and shown samples of various attacks and told what in them revealed their illegitimacy.

Following this, the pen test judged as more difficult for users to spot was administered to the users targeted in the previous tests. Using the protocol, users were sent the CRS a week after the test. Overall, 6 percent of the users were deemed high risk. The overall NRS was now 88 (i.e., 94–6). Even more encouraging, 4 percent of the users reported the attack, and there were just as many individuals from high- and low-risk groups reporting it.

Based on the success of the process, Masonry Inc. is now deploying the CRS across all its branches. The net reporting rate among users continues to be high, and the overall risk levels continue to decline. Starting with a net resilience score of 24 and improving to 88 showed the measured impact of the interventions. The best part is that David Marcus could explain the score to everyone in top management and to users in the organization. Another plus is that he also now knew for certain how vulnerable the organization was, who in the organization was at risk, and why they were at risk. Rather than training everyone all the time, he is now training only those who need it, in the manner that suits their need.

## CYBERSANITIZING THE HOSPITAL

Alex Sanchez, MD, MBA, was the CISO of a 450-bed hospital, Ashmore Medical Center (AMC), in Connecticut. The hospital employed 2,500 doctors, nurses, and support staff across multiple specialties. Sanchez was of the rare breed of older physicians who were also very technology savvy. He was an early adopter of many technologies, not just for his specialty but also for computing. He had been among the first in the hospital to adopt personal digital assistants (PDAs) for prescriptions, and this was back in the late 1990s, long before there were smartphones. He had developed databases, assembled virtual medical teams, and taken to telemedicine, electronic medical and health records (EMRs), and e-prescribing years before they became mainstream. His thirst for technology found him taking a sabbatical from the hospital to work for an EMR system developer for a while. Along the way, he earned an informatics degree, various security certificates, and the attention of the hospital leadership—who entrusted him with "all things IT." This came with many titles; some called him the CISO, but he called himself the chief informatics officer—a person in charge of the people and the processes they perform using technology.

Being in health care brought regulations, and this motivated many of the operations of Sanchez's team. Although there were many mandated IT protections, they were known, which meant his IT team of 18 full-time and another dozen part-time workers could plan for them. The unknowns were his primary worry. One was ransomware, which was crippling hospitals all over the country. Hospitals from Los Angeles to Louisiana were being hit and losing information; more recently, they had even hit a major hospital in New York. Sanchez knew that any day his hospital could be targeted, so he was preparing for it, creating backups of systems and looking for every weakness that hackers could exploit. But he knew that even the most well-prepared hospitals had problems coming back online—because many health IT systems were old, being kept in place to support legacy software, so a shutdown wasn't just about data recovery. Many of the systems had data that weren't capable of being backed up in the manner modern systems permitted. The quality of data they held, the liability of losing it, and the incalculable detriment of a long shutdown to the credibility of the hospital worried Sanchez.

He did his rounds, spoke to different physicians and caregivers working at AMC, and quickly recognized that the problem required more than just technical fixes; users needed to become better at recognizing social engineering attacks. The case he made to the hospital's executive leadership was this: "New, shiny cars aren't any less accident prone than older cars in the hands of a careless driver. We need driver's ed so everyone is better at driving. We need cyber hygiene training."

"But train them on what? To do what?," they asked. "How much would it cost? Why not just create better backups and be done with it?" And "what about resistance and downtimes? Is it all really worth it?" Having worked in the system for years, Sanchez knew that doing anything new with technology in a hospital wasn't easy. Everyone liked the idea of technology, but no one wanted to learn something new, especially when it had nothing to do with their core practice. Sanchez had learned this the hard way with EMRs: "Tell them about a new CT scanner and many of them will listen; tell them about how to prescribe online, and they would leave the room." Initially, physicians, nurses, and other staff detested EMRs—everyone except insurance companies, who loved them because they made the hospitals accountable. There was a reason the US had lagged behind the rest of the developed world in health IT adoption. It was because no one other than insurers wanted it.

Sanchez's team of physicians was among the first in the region to deploy EMRs—and successfully, too. It was no small feat, but this experience had taught him a thing or two about physicians. They liked evidence. They liked to see data, hard factual support. Many were trained in "evidence-based medicine" and were very receptive to a similar approach for health IT. So, Sanchez worked on building the evidence on EMRs.

He created a working subunit of EMR users within the hospital. From them, he collected data on how much individual time was spent inputting data into the system, a perceived cost, and compared it against the reduced lag time for reimbursement due to the system being online, a benefit. He tracked the time it took various caregivers in the unit to learn the system and even provided actual data on their patient satisfaction levels. In addition, he set up a sort of technology triage team that helped individual physician and nursing teams with their setup and questions. This was followed up with individualized training that was given based on a survey of technology skills and proficiency. The training, the help desk, the cost-benefit analysis, and even the final deployment of support were implemented gradually, each based on data that was procured, all of which was shared with everyone in the hospital. In short order, the overall resistance to EMRs was reduced; in fact, various subspecialties were asking for it. Within a year and a half, the hospital-wide system was up and running, and the overall satisfaction level among employees—another factor Sanchez accounted for—averaged a whopping 5 out of a possible 5. AMC became the poster child for how to successfully implement EMRs. They appeared in research studies and quickly became part of AMC's differentiating factor. You could come to AMC, where everything could be handled online—and this was all because of Alex Sanchez.

However, the same technology was now posing a new threat. Back before EMRs, everyone enjoyed doing things on paper, but ransomware could hurt that system. Now, with health IT, one rogue email opened by an intake person could corrupt the system—the same system Sanchez had implemented—but there was no turning back the clock. Not only were EMRs here to stay, but now everyone needed them. It was no longer just whether you had computing that dictated patient satisfaction levels but now also how much more accessible the caregivers were to the patient. There was no avoiding that Sanchez had to protect the networks, computers, devices, and the people who were using them. It had to be done for the sake of the patients who trust the hospital and for the future of the hospital.

The question was how best to accomplish this. There was a steady stream of new vendors offering various technological and nontechnological solutions. Each brought various short- and long-term lock-ins that came from licenses and training. These weren't minor—because each involved training almost everyone who worked in AMC, which meant taking time away from the core responsibility of caring for patients—and none provided the type of evidence Sanchez needed to convince everyone. Just as it had worked with EMR, an evidence-based process was necessary.

Sanchez already had a pen-testing program in place, where IT staff members were sending phishing emails. This program was fine, but the data from it appeared inadequate. Beyond pointing out who was falling for a phish, there wasn't much more that could be gleaned from it. The data was insufficient to convince users. Sanchez had read about my work on the Cyber Risk Survey, watched my Black Hat talks on the V-Triad and the risk profile, and read my research papers on the Cyber Hygiene Inventory (CHI). Altogether, they sounded like the answer to his problem. With a combination of the CRS and CHI, he could have all the evidence he needed.

We began by conducting a baseline cyber hygiene assessment using the CHI. Sanchez measured the four facets of cyber hygiene (awareness, knowledge, capacity, and enactment) across the SAFETY dimensions (storage and device hygiene, authentication and credential hygiene, friends and social media hygiene, email and messaging hygiene, and transmission hygiene). The survey was deployed via email to a sample of two thousand employees, stratified into four groups—physicians, nurses, administration/management, and general staff—based on their overall proportions. The survey data was collected over eight weeks, with personal emails from Sanchez to respondents and even some in-person visits to help anyone too busy to respond.

Next, a phishing test was developed using the V-Triad and the Cyber Risk Survey. Using the design steps detailed in chapter 7, we developed four different phishing tests. Each had at least one element from each vertex of the triad and followed a development process similar to that outlined in the previous case. The end product of the process was two phishing tests that were ready for deployment.

One was a spear phishing email with a hyperlink to a pathology pickup notice. It mimicked a specialized service that was used by hospitals in the region for collecting biological samples and borrowed some fonts and colors from their notification emails. The subject line provided a pickup time and asked the user to click on the hyperlink to read the special package preparation instructions. The second email was made to appear as an ongoing exchange with a patient who had attached their health test results. It included the word "Re" followed by "as discussed, test results" in the subject line of the email. The body of the email addressed a generic nursing staff and asked the reader to open the hyperlink. A baseline test of 35 full- and part-time IT staff and another 15 physicians and nursing staff members who were part of the technology advisory council netted a score of 9 on the CRS for the delivery email and a score of 5 for the patient communication email. The reason for the discrepant results was that a pathology pickup email would likely be no different from other phishing tests that users had already been trained on. Given its difficulty rating, the patient email was found to be more appropriate and was used for the first test.

Because of the in-house phish-testing program in place, it was easy to deploy the phishing email using the existing process that the IT team had developed. The test was deployed to a sample of two thousand respondents and followed many of the CRS protocols detailed in chapter 8 (and laid out in the preceding case study example).

Overall, based on the response to the CRS, 625 respondents fell in the high-risk category, for an NRS of 37.50 (that is, 68.75–31.25). The answers to the CRS allowed Sanchez to understand why the users who were high risk ended up falling for the phish. For the sake of brevity, I am glossing over the findings of the open-ended analysis of the CRS, but suffice to say that just like David Marcus in the previous case, Alex Sanchez could analyze why users fell for the phishing attack.

Next, using the response to the CHI, Sanchez conducted a gap analysis. This basically involves comparing the differences in levels of awareness, knowledge, enactment, and capacity across the SAFETY dimension of the CHI. On top of this, Sanchez overlaid the CRS-based high-risk versus low-risk tiers. The chart in figure 9.1 shows the overall gaps.

Figure 9.1 displays the gaps between average awareness and enactment levels among high- and low-risk users for each cyber hygiene dimension (where the responses were measured using a 0–5 response scale and averaged). Let's examine each of them. Beginning from the left, there was storage hygiene where the high-risk users scored low on both awareness and enactment compared to low-risk users. In the next factor, authentication hygiene, an issue of importance for protecting against social engineering, Sanchez found even lower levels of awareness among high-risk users but the gap with enactment was around 1, roughly the same as in the previous dimension. The friends and social media hygiene had relatively high awareness and enactment in both groups, but here the gaps among users were the same. The next important factor from a social engineering point of view, email hygiene, showed that low-risk users had lower overall awareness and

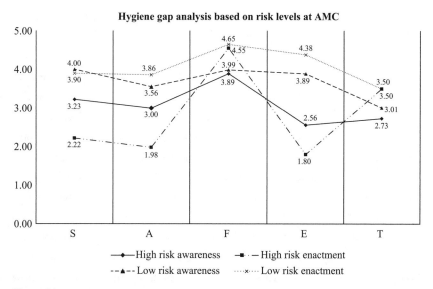

**Figure 9.1**

Results of AMC's cyber hygiene gap analysis.

*Note*: The x-axis includes each of the five CHI dimensions, and the y-axis represents the aggregated response of employees.

very low enactment hygiene (average enactment gap of 0.76) compared to high-risk users (gap of 0.49). Finally, both high- and low-risk users scored relatively high in transmission hygiene, with a little gap in enactment.

With this data, Sanchez obtained an understanding of the core of the issues. There were relatively significant gaps in awareness and enactment when it came to high-risk users. These gaps differed based on the hygiene factor and were considerable when it came to email and authentication hygiene; for the former, the overall awareness itself was low, as was enactment.

The analysis was just the tip of the information trove that the combined use of the CRS and CHI netted. There were individual questions in the CHI that helped Sanchez drill down to where the gaps were greatest. There were also the open-ended questions from the CRS that he could map onto the hygiene gaps, which he could use to better define the problem areas and, by examining the responses of low-risk users, identify solutions. Thus, using the CRS, he recognized which users were likely to fall for a ransomware attack, and, using the CHI, he knew what he needed to do about it.

The analysis went further and examined capacity constraints, which allowed Sanchez to identify what the users felt they lacked. For instance, the data showed that there were no password vaults being provided, no official VPN service being offered, no individual device backup system, no help desk for forwarding questionable emails that might violate a health regulation, let alone one for reporting phishing attempts, and no uniform communication platform for sharing such information. The most important part of it all was the net resilience score. This was easy to compute and easier to understand. Sanchez now had evidence that could help explain to the employees of the organization why changes were being made, who it was going to affect, and when.

The NRS along with the cyber hygiene inventory data allowed Sanchez to seek and identify vendors who could address specific issues. He could make a case for capital expenses to the hospital administrator and show them how the hospital would benefit. He now had a process in place to check and assess overall readiness. He had a score for it, a target score he needed to get to, and a timeline and action plan for doing it.

Using the CHS and CHI, Sanchez could measure, assess, track, communicate, calibrate, and, most importantly, convince. He could check to see what was necessary, what types of interventions would work, and who needed them. He could show who needed training and why, and he could build accountability. Sanchez could finally build his evidence-based cyber hygiene model and, without a doubt, cybersanitize Ashmore Medical Center.

## THE TARIFF ADVISORY COMMISSION'S COMMUNICATION POLICY

Elaine Banks, CISO of the Tariff Advisory Commission (TAC), had instituted many user-focused initiatives, from training to pen testing and more. Some were mandated requirements, whereas others were implemented because she felt they were necessary. Her charge was relatively manageable. Based in Washington, D.C., TAC was a government agency with 500 employees, small compared to other government organizations, where this number of employees would be a mere department. But given the small area she covered, she wanted to ensure that every person in the organization was resilient. "Nothing like having a small agency make the news for being breached. . . . You end up becoming the poster child for cyber vulnerability," she said.

Banks knew that some solutions worked better than others, but which ones? There was training that everyone complained about but that had, after two years of implementation at TAC, netted less than a 1 percent failure rate. Then there was email marking, a simple protocol for highlighting which emails were external, which no one complained about but that appeared to be useful. But which of these mattered, and how much? Such questions led Banks to apply the CRS protocol.

As part of it, we developed a phishing pen-test email using the V-Triad. The email changed the "external" flag into an "internal" flag and said the "email being internal should be trusted." This reversed the usual flag that warned users to suspect an external email's links. This satisfied the V-Triad's *compatibility* vertex. To make the email *credible*, we mimicked the internal calendar system and sent out a meeting request for reviewing material that was attached to the email as a PDF. Finally, each email was *customized* by including the names of a handful of other individuals (of course, none were actual employees) on the invite list, so it appeared that others in the organization were also being invited.

Given the small size of TAC's IT department, the CRS benchmark was provided by five IT staff members. The CRS benchmark netted an 8 on the CRI, showing that the attack should be rather easy for users to detect. The major reason was the obvious nature of the compatibility signal "internal," which almost everyone was expected to notice.

The CRS deployment after the attack resulted in close to a 97 percent response rate, not surprising given the small size of the organization's user population. From those who had recalled receiving the email, about 18 percent of the respondents were scored as high risk—far above the less than 2–3 percent pen-test failure rates the organization had experienced in the past year. The NRS was 64 (i.e., 82–18).

The diagnosis of the reasons for their risk pointed to two major issues: heuristics and risk beliefs. Many users were looking at the subject line as a new indicator of internal emails. They reasoned that IT managers might have changed this policy to highlight internal emails as well and that they had missed the announcement about the change. Within the body of the email, most noticed the calendar invitation, merely glanced at the list of other names on it, and believed them to be someone else in the organization they weren't familiar with. While most of the names were phony, that there *were* other names, they reasoned, made the invitation appear authentic. As one user wrote on the CRS: "I figured I can't know everyone in the organization and there were at least a few familiar looking names. If not me, someone from that list would surely have informed others of it being fake!"

The ensuing analysis helped Elaine Banks understand more than all the other tests she had deployed. She knew where her training efforts needed to be focused. Rather than simply keep using the old form of training, she focused on improving heuristics. But that wasn't the only issue. Thanks to the CRS, she realized there was a sizable gap in the organization's IT communication practices. Users learned about policy from other users and relied on others for guidance. The proverbial blind were leading the blind because the IT division responsible for providing the light wasn't doing so. The other pivotal piece of information was the NRS, which allowed Banks to understand how resilient her overall organization was and to set a target score for the future.

To achieve this, Banks worked with the IT staff and developed a comprehensive communication plan that outlined how IT policy would be implemented and how users would be informed about it. This communication plan made it clear how changes would be made and how users would be informed about them. Banks and the IT staff also came up with a rather novel way of communicating IT policy. The basis for this was a response from a low-risk user: "Like they do in hotel rooms and our office building, where they visually present the fire-safety and evacuation instructions, why not provide a visual roadmap of IT policies and expectation?" So, rather than place posters asking users to be careful about phishing, Banks had the IT department develop policy posters that showed how it would communicate with them about policy. This included mention of the specific website where IT changes would be announced and a help-desk email and phone number that could be used to ask for information. It also showed the steps users should take if they saw something suspicious online—what to do, who to contact, when, and how. These were simple solutions, though Banks had never considered them in the past. Within weeks, reports of suspicious online activities were pouring in to the help desk. Everyone was giving suggestions in the newly developed online portal: suggestions for improving heuristics, best practices, and solutions. Users were clued in, vested, and engaged in cyber resilience like never before. Banks was sold on the value of the CRS.

## CRAFTING A SIMPLE SOLUTION

Amy Craft was like many other CISOs in that she wanted not just to improve the overall cyber posture of wealth management company Amazing Income Inc. but also ensure there was never a breach—at least never because of something avoidable, such as a compromised user's credentials being used.

Amazing Income was a dynamic young company that was growing very rapidly. In less than a decade, it had grown from a single small office of 12 employees to 550 employees in 20 regional offices. Its growth occurred because it was agile, young, dynamic, and excelled in meeting the customer where they were—at home, at work, at a cafe, even at an airport before they flew out of town. Its regional offices functioned as semiautonomous small businesses.

Each office used to have its own IT procurement systems, approved devices, and support protocols, but Craft had changed all that. She centralized IT support services and started crafting a company-wide IT policy that specified not just what users could or couldn't do on their devices but also what devices they could do it on.

Craft's biggest fear was the employees in the organization—they were the organization's most valuable resource but also the biggest weakness from an IT security point of view. While she could centralize a lot of IT actions, she couldn't control what users did on their devices. She knew it was likely that a breach would occur someday. The nightmare scenarios were endless—ransomware, credential misuse, theft of data, loss or breach—and all because of social engineering.

To protect against this, Craft had instituted several new technical safe-guards, but she knew it wasn't enough because the users were as yet untrained. Every other CISO she knew was also training their users. Many agreed that users were like cats and that training wasn't making them perfect but it was better than having them run wild.

After instituting a company-wide training program that took an entire month to cover all branches, Craft wanted to know what more was needed. "Money had been spent, mandates met, time lost, but what about readiness? Are we there yet?," asked company leadership. "And how much more do we have to spend to get to that . . . whatever that state is?"

While everyone at the top wanted to train everyone and be cybersecure—no matter what it took from Craft—every team manager complained about training and felt it was a waste of time. "We all know better not to open some emails . . . plus, you are wasting my time and that of my branch staff where instead of bringing in clients, they are working on some training on using email safely!," remarked a younger branch head. So, Craft wanted to know who needed training, what they needed, and whether something other than training might instead suffice. She also wanted to know where Amazing Income stood as an organization and how exposed they were to social engineering.

We began developing phishing tests using the V-Triad to assess the users' readiness. This process led to four pen tests for the benchmark that were scored by a group of 25 IT staff members. Based on their assessment, we picked two tests, one that scored a benchmark of 6 and another that scored a 5. Both were thought to be on a par with each other in terms of difficulty.

The first was deployed to 330 employees in the beginning of the first quarter, followed by the CRS assessment, which occurred a week after the test. After removing respondents who did not recall the test, the net response rate was 80 percent. The final CRS showed 33 percent of users as being high risk, for an NRS of 34 (i.e., 67–33).

The analysis suggested that a preponderance of high-risk users were suffering from a lack of good heuristics, poor risk beliefs, and poorer habits. Stratifying the high risk into moderate and highest risk (by rank ordering users and taking the median) suggested that those with the highest risk were primarily suffering from poor risk beliefs and bad habits.

A habit intervention and a program to improve risk beliefs were clearly needed. Based on the diagnosis, a mindfulness-focused intervention along with a training program aimed at improving risk beliefs and heuristics, like the one used at Masonry Inc., was clearly needed and was recommended.

But there was a diamond hidden in the responses of the low-risk users. Some users in the lowest risk levels had come up with interesting email use rituals and practices. For instance, one user had created separate in-boxes and set up mail delivery rules such that IT emails showed up in a specific in-box, all client emails in a separate folder, and emails from all immediate staff in another folder. These had custom icons and colors. Another user had also created a process of her own. She had set up rules that allowed only certain interdepartmental emails to show up during work hours, based on their priority level. Every staff member under her supervision was trained to use the priority system, and without that emails were not being attended to and being kept in different priority-based folders. Most users in that division were scoring lower risk levels and had fewer issues with habits than those in the divisions that did not institute such workarounds.

Craft used these findings to design an organization-wide system of email delivery rules and a system that flagged emails based on their importance, which everyone in the organization was trained to use. The system was launched on a limited basis in half the branches in the organization, which allowed the second pen test, conducted the following quarter, to assess its efficacy.

The second pen test netted a significantly lower percentage of vulnerable users. The overall percentage of higher-risk users was 18 percent, and although the NRS of 64 (i.e., 82–18) was still higher than acceptable, this was a rather remarkable improvement in just a quarter. The risks from poor habits were significantly lower in the groups using the new email rule system. Even more encouragingly, overall, heuristics-based failures had also been reduced significantly, and risk beliefs across all users had improved. Best of all, Craft now better understood her users. Rather than just focusing on training, she began conducting interviews and seminars where low-risk users would meet and tell higher-risk users their tricks and techniques for improving the organization's overall cyber resilience.

## FROM DISMAL TO DEFENDED

Akron House is a small nonprofit serving some of the poorest neighborhoods of northern Pennsylvania. The organization is headed by its inspirational leader, Jake Traposh, a community activist who felt a calling to help the poorer residents of the region. The organization serves a diverse population of the extremely poor, those who had fallen through the cracks, and poorer new immigrant refugees, chronically homeless people, and senior citizens.

The nonprofit ran senior facilities, halfway homes, and even a rent-free rooming house, soup kitchen, health care clinics, and daycare facilities. Traposh's staff of 45 was involved in all aspects of managing these services.

They were generalists, as many in nonprofits, with their thin budgets, are. Other than hiring a dedicated accounting person, there just weren't enough resources to hire specialists to manage marketing, let alone cybersecurity. Everyone had to pitch in and do anything when necessary. This also meant that almost everyone had to have access to all the files and folders that Akron House handled. This included highly sensitive personal, clinical, and criminal histories of all their clients, and data from funders and other reports that the staff was routinely generating. Except for internal salary data, nothing was off-limits to anyone. This worried Traposh, especially after he heard about the many ransomware attacks that were targeting organizations in the region. He wasn't worried about the personal data alone; it was all the files, folders, and historical information that they had accumulated that was pivotal to their clients, many of whom relied on Akron House to keep it safe. Even with backups, having these hacked, especially given the vulnerability of the population Traposh served, could be catastrophic. He wanted a system, some specific mechanism that would help protect his staff and identify who among his staff he had to worry the most about. He wanted to know his weak links.

Traposh hadn't spent much on cybersecurity awareness training. Other than sending users some online educational material, he hadn't done much else, so the need for him to employ dedicated security training was obvious. But he was also interested in understanding who needed it and what types of training they would benefit from the most. He also wanted simple solutions, which he could employ right away, that would help protect his organization before even the training kicked in.

Working with him, I developed two phishing pen tests and, using the CRS approach, ascertained the benchmark for each test. The benchmark was provided by one user, Jake Traposh himself, and I selected a test with a moderate level of difficulty. Rather than deploying the test just once, it was deployed on three nonconsecutive days over a period of a week.

Following the CRS protocol, the survey responses were collected within a week after the last attack. Almost everyone targeted responded to the CRS. Forty percent of them had fallen victim to the attack, for an NRS of 20 (i.e., 60–40). None of this was surprising, given the small size of the organization and the lack of any training. What was surprising, however, was that at least 10 of the users had clicked on the phishing link multiple times.

In examining the responses, the overall problem was not just the lack of adequate knowledge, which led to poor risk beliefs and even poorer heuristics, but also the preponderance of dismal habits. Bad email habits were markedly higher among the high-risk users, and they stemmed from their use of unified in-boxes, where they had populated multiple personal email accounts along with their work accounts. This increased the number of emails that would appear in their in-box. Many users were using email clients that allowed postponement of emails, such that the emails would reappear at a different time. Such actions were contributing to the repeat appearance of emails in in-boxes and with that the chances of victimization.

We first dealt with the lack of good heuristics and risk beliefs by embarking on a training campaign that involved teaching everyone simple rules that they could remember about online actions. These were posted at various locations throughout the organization. Adding to this, Traposh hired a training organization that trained users about safe data handling and email use.

Besides this, Akron House used the CRS-based risk assessment tiers and mapped them onto the organizational chart, where Traposh interpolated what each user was functionally doing. This helped define data limits and separate out who needed access to different organizational material and who didn't. Next, we removed access to email from all the internal computers. Instead, everyone was given Apple's branded tablets and phones, on which email was accessed. This created a simple air gap that made it harder for a social engineering attack to cripple the entire organization.

In the months following the attack, we conducted a second test, which showed a marked decrease in clickthroughs, with only 8 percent of the users being high risk. The NRS was a whopping 84 (i.e., 92–8). Multiple clicks ceased completely, and the overall riskiness decreased. All the clicks were contained within mobile devices. Although some issues with risk beliefs persisted, these were addressed by following approaches similar to those discussed in the previous two vignettes, using a combination of technical interventions and personalized solutions. A third attack, which was far more sophisticated, followed at the end of the calendar year. This netted even fewer victims than the second one, which, considering the increased difficulty of the attack and the short time frame, was remarkable. Better yet, the victims had clicked on the link on a mobile device where the attack's potential impact was contained. Thanks to these simple interventions, the organization's risk posture had gone from dismal to defended.

Each vignette shows how you could apply an intervention to improve user cyber resilience in a pointed, cost-effective, evidence-based, and focused manner. All this was possible only because of the CRS and its associated metrics. But these aren't best practices. They merely show what is possible. Of course, even more is possible if you implement the CRS, and by this point in the book, you are ready to do this.

# 10  REVERSING THE SOCIAL ENGINEER'S ADVANTAGES

There is the parable of a man who late one night was frantically searching for his car keys under a streetlight when a police officer stopped to help and asked where he'd lost the keys. The man responded, "Over there, in the alleyway, by my car." "But then, why are you looking for it here?," the officer asked. The man replied, "because the light's better here."[1]

His response indicates "the streetlight effect"—a type of observational bias where people look for solutions where it is easy to do so instead of where the problem really exists. It is a fitting analogy for how the cybersecurity community has tried to solve the problem of social engineering. Rather than attempt to understand users—the primary target of the attacks—they have applied training solutions because it is what engineers understand and have available to them.

Such thinking has made companies in the cybersecurity engineering and solutions space enormously wealthy, but, as this book has detailed, IT managers have reaped no benefits. They face even greater pressure from organizational leadership wanting absolute security and from insurers wanting complete accountability because everyone believes the problem has an engineered solution that simply needs to be better implemented. Adding to this narrative are marketing materials from security consultants, advocacy from policy makers, and guidelines from a variety of federal, state, and international government agencies. All call for more training without specifying how, when, and how often it is needed. Now, over six years after the Sony Pictures Entertainment (SPE) breach that started the cybersecurity awareness training bandwagon, security training hours per user have ballooned. Almost every major organization in the world engages in user awareness training, but we are no closer to stopping social engineering attacks from wreaking havoc.

In fact, in the ensuing years, such attacks have become even more successful. A few months after the SPE attack, in the aftermath of a ransomware attack on a California hospital, I again called for studying users, the weakest links in IT, who are the targets of such attacks. Otherwise, I predicted, 2016 would likely see many more attacks—it could be "the year of online extortion."[2] Most in cybersecurity ignored my warning, but I was wrong on one count as well: it wasn't only that year; it has been every year since.

The trend continued well into 2019, and 2020, when city governments all across the US Midwest and South were crippled by ransomware attacks on well-known technology companies such as the GPS provider Garmin, which suffered attacks that took down its fitness products and services. The attackers demanded $10 million in ransom, which reports suggest was paid by Garmin. Similar payments were made by many top universities in the US, including Columbia College, the University of Chicago, Michigan State University, and the University of California. This continued in 2021, with even bigger attacks. There were attacks that crippled the world's largest meat processor, JBS, which paid $11 million, and Colonial Pipeline, which supplies petroleum to half the US East Coast, which also paid the ransom. Altogether, they speak to how, even after applying training and all the many containment and constraining solutions discussed in chapter 3, organizations are helpless against social engineering.[3]

This chapter will discuss why these trends are worsening. We'll examine the long-term changes in the macro environment, which have altered the demand for hackers. We'll also discuss how the "streetlight effect" and a bias toward sophistication are contributing to the problem. Finally, we'll see how the Cyber Risk Survey helps correct and tilt the advantage back in favor of IT managers.

### THE ADVANTAGE TILT

In chapter 2, we discussed four advantages that social engineers have over organizational IT managers. To recap, hackers have an *informational advantage* that comes from asymmetric information available to them about users. This comes from the troves of personal information on users already available on the Dark Web, and much more being added as breaches and hacks ceaselessly take place. Hackers' *knowledge advantage* comes from the low effort required to craft compelling phishing attacks. It also comes from the availability of technical resources, including coders and malware-for-hire. With this, even amateur hackers are able to craft highly sophisticated attacks.

Furthering these advantages is the prevailing *information environment* in which organizational IT operates. Because of varying notification laws and breach-associated liability, most organizations maintain proverbial IT security silos where everyone maintains operational silence. They learn from their internal trials and mistakes and make little actionable information available for others. Consequently, every organization experiences a steep learning curve in protecting against each attack.

The *incentive framework* makes things worse. Organizations are castigated for being breached, their IT departments and CEOs are publicly humiliated, and many lose their jobs. This motivates secrecy and risk aversion. It prompts organizations to comply with hacker demands, even paying them off, and it promotes a cover-your-ass (CYA) culture,[4] where a solution is implemented simply because a federal agency is recommending it or because major organizations are applying it, even if the solution lacks efficacy. This way, in the event of a breach, the agency's recommendation or the solution's prior use can be shown as proof of effort.

In contrast, just like graffiti artists showing off their designs, hackers are willing, even eager, to publicize their exploits. It brings them more notoriety, more credibility, and more clients willing to pay for their skills. Hackers keep trying different attack types because they incur no cost or ramifications. They trade their wares in specialized marketplaces and on encrypted platforms using hard-to-trace digital currency. Hackers learn, share, collaborate, adapt, and prosper. Because of *information* and *knowledge advantages* and because the *information environment* and *incentive framework* favor them, much like the Thuggees we met in chapter 1, today's hackers succeed in virtual robbery.

But the future of cybersecurity looks worse. We haven't fixed the fundamentally flawed password-based authentication system, which (as discussed in chapter 2) is why social engineering began and thrives. We haven't come to terms with the fact that security training, as we do it now, doesn't work, and we lack any other universally applicable, proactive solution to protect users from social engineering. This has attracted more entities to spear phishing. More investments, innovations, and newer talent are flooding the hacker space. Adding to this are changes in the marketplace. The advantages are tipping even more in favor of social engineering.

For one thing, every government in the world now recognizes the value of social engineering. They see its potential for attacking dissidents and opponents. What were in the past mostly operations by rogue nations such as North Korea and then by hackers backed by the Russian Kremlin, the Chinese Communist Party, Iran, and Syria are now being conducted by almost every major nation. Each has invested enormously in social engineering, some even devoting entire espionage units to such operations.

This has supported the rise of for-profit hacking services such as Israeli private security company NSO Group Technologies, Nigeria-based Black Axe Syndicate, and Italy-based Hacking Team. Each conducts highly sophisticated social engineering attacks for various government-backed entities. Because of this, anyone can be targeted and hacked.

The most glaring evidence of this is the social engineering attack launched by the Saudi Arabian government of Mohammed Bin Salman on Jeff Bezos. It was ostensibly conducted by NSO using Pegasus, a potent malware that was capable of accessing a smartphone's microphone and camera and reading every bit of data on it. Pegasus is a zero-click exploit, which can infect any smartphone via messaging, even if the user does not click or open the message. Using it, all data from Bezos's device was exfiltrated and used to taunt and influence him. This only came to light when Bezos went public in the media.[5] It shows that even the richest man on the planet, with all the resources available to him, cannot protect himself from social engineering.

Of course, Mohammed Bin Salman wasn't the only head of state using NSO's tools. In 2021, an international investigative journalism initiative found that Pegasus had been used by political parties and government agencies in different parts of the world to spy on over 50,000 journalists, opposition political parties, activists, and businesspeople.

This shows how nations' involvement has altered the motivations behind social engineering. What earlier had been driven more by monetary interest is now motivated by maintaining sovereignty and achieving national pride. This has tightened the nexus between technology developers and hackers and created a rich ecosystem of skills and suppliers who are protected by the state. Another example of is the 2019 security warning about the Chinese military creating backdoors into equipment provided by Huawei, one of the world's largest suppliers of telecom networking kits, which operates out of China.[6] Had this gone undetected, it would have provided the Chinese military with access to mobile communications in much of the world.

Cases such as this reveal the sheer scale of investment and the high stakes in social engineering. There are enormous resources available for developing deep exploits, high demand for hacking capabilities, and a ready market for hacker services—which inspires many more hackers to develop spear phishing and pretexting attacks.

A second factor tipping the scales in favor of social engineering is the emergence of a few global market leaders in the technology product space. As I discussed in chapter 2, having a few dominant systems makes it far more lucrative to invest resources in developing exploits for them. For instance, by developing a ransomware that could hack into Windows machines through a pop-up advertisement, for years a social engineering ring in New Delhi, India, used a dozen telephone call centers to fleece computer users all over the world out of millions of dollars. They spent time and effort developing just that one exploit because the system could be reused globally. This ransomware ring wasn't detected by Microsoft. It was stopped in March 2020, when an undercover investigative report by BBC News exposed it.[7] Exploits such as these are likely to persist, as the world's technology sectors appear to be coalescing on a handful of dominant global players (e.g., Facebook for social media, Amazon for retail, Google for search).

Such attacks are aided by the lack of universally applicable laws protecting user data or mandating security standards. With more and more hardware and software products being programmed or assembled in far-flung nations across the globe, the chances of rogue employees stealing data, foreign adversaries injecting malware, and organizations mining user data without regard to the laws of the nations where the products are actually being used have increased. This was the case with the multimillion-dollar IRS telephone scam, also perpetrated in India by another hacker group using telephone call centers (discussed in chapter 2). These scammers possessed confidential information about US consumers taken from India-based outsourcing vendors.[8]

Further abetting this is the lack of globally applicable breach notification standards. This means a breach of US consumer data housed in a developing nation might never come to light—making the consumer open to future exploits. So, when American consumer data is stolen from a back-office contractor in India, we may never hear about it in the US. Also, because of individual state mandates on breach notification, even if an organization in the US learns of that breach, they might never come clean about it to their customers. Such trends are likely to get worse as organizations continue to outsource coding, product development, back-office operations, and even managed security to lower-cost nations.

Each intermediary increases the number of devices through which data passes, which significantly enhances the potential points available for hacker exploits and the number of users with privileged access. This reduces the number of degrees separating the hacker from an authorized user—which we know from chapter 2 is a critical issue—and the likelihood of a successful breach.

A third factor tipping the scales in favor of social engineering is the ever-expanding capabilities of technology. This includes the harvesting of existing capabilities and new inventions that are changing the scope and scale of user data being captured. Consider cameras as a case in point. In the early 2010s, having cameras in both the front and back of the iPhone was revolutionary. Today, almost every gadget contains a camera or two. Tablets, smart streets, home security systems, cars, garage door openers, refrigerators, watches, and doorbells have multiple cameras and microphones. These capture user data, their pictures, videos, ambient sounds, and biometrics and store them on various devices, services, and servers. As we know, user data is the currency of the internet, and these existing technologies gather ever more granular data about users. They have significantly expanded the overall marketplace supply of user data and with it the demand for exploiting it.

Developments in quantum computing, machine learning (ML), and artificial intelligence (AI) are adding to the technological arsenal of hackers. Many are in early stages of development but are capable of transforming the threat landscape. For instance, quantum computing, which is already in an advanced stage of development, will make cryptographic protocols based on Turing systems obsolete. They will make it possible to hack into all existing computing machines regardless of passwords and encryption levels. For now, quantum computing continues to be cost-prohibitive and accessible only to deep-pocketed nations. But, as we have learned from the history of many technical exploits (such as NSA's EternalBlue, which was stolen and led to WannaCry in the hands of hackers), quantum computing, too, will eventually lead to future exploits by rogue nations and hacking groups.

In support of this trend is growing evidence that some hackers are using open-source AI and ML programs to personalize spear phishing attacks. Such attacks can change their persuasive appeal in real time based on each user's browsing habits and online data trails. Thus, it is simply a matter of time before hackers use quantum systems to crack passwords—forever changing the face of cybersecurity.[9]

The final factor tipping the scale in favor of the social engineer is the rise of newer organizational practices. These are being rushed into reality by environmental uncertainty. The COVID-19 pandemic is a telling example. It has demonstrated the importance of remote work for organizational continuity and survival and has ushered in virtual work practices throughout the world. Even in companies that traditionally didn't allow it (e.g., law firms and medical offices), users are now allowed to work from home on consumer-grade computing networks using residential internet access portals and devices, which are outside the ring-fenced security of organizational networks. Users are also uninhibitedly surrounded by internet of things (IoT) devices of their choosing, using shared devices, and performing cyber actions completely outside the purview of trained IT staff. IT management's control over users, which as we know was already eroding, is now almost completely ceded. Those who stood in the path of the social engineer are all but gone.

But while the trends point to the increasing potency of social engineering, its march can be checked. For this, we need to change the existing mind-set that blanket training or a technical "magic bullet" will solve the problem. But why do we continue to think in these terms? The next section explains why.

## THE SOPHISTICATION BIAS

Just after midnight on Saint Patrick's Day, March 18, 1990, two individuals entered the Isabella Stewart Gardner Museum in Boston, Massachusetts. In what became the biggest art heist in US history, they stole $500 million worth of priceless art that has never been recovered.

The thieves were after very specific artworks and knew exactly where they were displayed. The duo first went to the Dutch Room on the second floor and stole five famous works by Rembrandt, including a self-portrait, and then a masterpiece by Vermeer, one of only 36 works he painted in his entire life. They walked to another room at the end of the same floor and took five works by Degas and an eagle finial that adorned a flagpole. Along the way, they passed works by Michelangelo and Raphael and left them untouched. Finally, they went to the first floor and stole a piece by Manet.

The thieves knew the workings of the museum, the security systems, and the guards. They knew there were two guards at night, whom they locked in the basement. They knew how the infrared motion-detecting equipment and closed-circuit cameras worked. Before leaving, they took the videocassettes, destroyed the alarms, and removed the printouts from the motion-detecting equipment. They somehow even evaded the motion sensors on the first floor when stealing the Manet. The only reason we know any of this is because a backup of the infrared motion data was found on the computer's hard drive. There was nothing else: no fingerprints, no images, and no clues. Just some broken frames and shattered glass.

But was the heist sophisticated? At first blush, yes. It had to be. It was meticulously planned and executed. The museum was wired with state-of-the-art sensors and cameras and had guards on patrol. Somehow, the thieves managed to get past all this. They spent 81 minutes stealing—a lifetime for art thievery, which averages 5–7 minutes. They have also managed to evade arrest for over three decades. With artworks so famous, it is impossible to sell or display them without someone recognizing their provenance. Whoever perpetrated the heist had to know not just how to steal but also how to sell the art.

How did the thieves do it? By first ringing the doorbell. The thieves were dressed as Boston police officers and claimed to be checking on a reported disturbance. The security guard buzzed them in. According to Anthony Amore, the museum's chief of security and chief investigator, "Every other museum in Massachusetts with Rembrandt paintings had been robbed before."[10] The Gardner museum even knew their collection was a target. Yet, there were just two security guards protecting it that night. There was one buzzer to call the police, located at a guard's table. Once this guard walked away from it to open the door, he could no longer access it. The thieves overpowered him, asked him to call the other guard doing rounds, and handcuffed them both. They then had the run of the museum. The entire theft was possible because the guard believed it was really the police and let the thieves in. The single biggest art heist in US history was the result of simple social engineering.

But this isn't obvious unless you focus on the details. Our inclination is to think of the complexity of the heist: how it was planned and executed and what happened to the art. This causes a sophistication bias, where the complexity of a problem, event, or situation overwhelms our sense of reasoning. At times, it is deliberate, because of how a situation is presented by investigators or by the media. At others, something in the event draws us into its complexity, making it hard to see past that. The outcome of this is a bias away from the simple, and it happens more often in cybersecurity.

Let's look at what's now being called the biggest cyber breach of this century. On December 8, 2020, cybersecurity firm FireEye announced that the tools it uses for conducting penetration tests in thousands of organizations worldwide had been stolen by a Russian state-sponsored hacking operation. When investigating the theft, they discovered the conduit the hackers had used. It was a network monitoring software program called Orion, developed by SolarWinds. Orion was used by 33,000 other organizations, including the US Treasury Department, State Department, Department of Commerce, Department of Energy, and 250 other federal government agencies and defense contractors. It was also used by NATO, the European parliament, Microsoft, Intel, Nvidia, Cisco, Belkin, and VMWare.[11]

Somehow, the hackers had penetrated Orion's software and injected malware into its core build. It then made its way into user networks as part of the Orion software update. Because the malware was now part of the official software, anything the software could access became accessible to the hackers. For instance, because Orion was integrated with Microsoft Office 365, the hackers could access all emails and Office documents.

The malware was also very well designed. It stayed dormant for days on end before installing additional exploitation tools that allowed the hackers to move in the network. It searched certificates within individual devices and procured official access to cloud services where files were stored. The malware then encrypted and exfiltrated data in small amounts to avoid detection. It even stored the files on US-based servers such as Amazon's AWS so it could evade the more sophisticated early warning systems developed by the Department of Homeland Security that tracked the eventual destination of files.

This helped the attack evade detection for nine months. While we still don't know—and may never know—the true scale of what was taken, the breach is thought to have affected 18,000 organizations. Microsoft conducted an analysis of the breach and concluded that this was the most sophisticated attack of all time, requiring the efforts of at least a thousand software engineers for development.[12]

But how sophisticated was it? Sure, coding the malware required a lot of skill and capability, but even coding the Sony Pictures Entertainment (SPE) malware, which we discussed in the Introduction, required coding skills that no one at Sony could match. Besides, the Russian hackers involved in the SolarWinds attack are thought to be the same group that breached the DNC in 2016. As you may recall from chapter 2, they used a 14-year-old Canadian teen to conduct spear phishing attacks. We have yet to know whether this is the strategy they used again. After all, it had worked for them in the past. But from what we know, they probably didn't even need to. SolarWinds's security practices were already lax.

According to one analyst, the password to one of its servers was "solarwinds123" and was publicly visible because of a network misconfiguration. Until recently, because of problems with Orion's original software code, they had even asked their clients to disable virus scanning. These are some of the simplest cyber hygiene actions we ask of all users.[13] Yet, SolarWinds, whose clients include some of the most secure intelligence services in the world, failed to follow them.

Like the Gardner Museum, SolarWinds made the attack easy. They literally left the door open even though they knew they could be targeted. Yet, the media's focus was on the sophistication of the malware, on how many coders were used by the Russians, and how we could improve network monitoring and software coding to avoid such attacks. It was the sophistication bias playing out. It happened to some extent after the SPE hack, after the Office of Personnel Management breach, and after many others.

The sophistication bias appears endemic to cybersecurity for five reasons. The first reason is that breaches occur on computing technologies, which are already seen as potent. Thanks to computing technology, we have created efficiencies like never before, formed industries that never existed, and solved problems that were intractable. We have sent rovers around the universe, managed people from across the world, and communicated seamlessly with everyone. The potency of computing is undeniable. It is apparent and ubiquitous.

But this wasn't always obvious. Less than 25 years ago, business school textbooks were convincing would-be managers to use computers. The education system and the media were pushing people to recognize the computer's capabilities. They were selling us on the value of adoption, on its prowess. Those who brought the newest, shiniest, and latest in technology were called innovators. Those who didn't were labeled Luddites and laggards. Because of such influences, the idea that technology was potent became part of our collective consciousness. With it came a reverence for it.

A second reason the sophistication bias is common in cybersecurity is that computing technology is inherently complex. At its core, it involves complex mathematical operations and computer programs written in languages that machines interpret. Developers require more than a basic understanding of programming languages. They require knowledge of algorithms, big data operations, cryptography, and distributed computing. These are difficult for most people to understand, let alone explain.

Take the case of a technology such as blockchain. The basic blockchain technology involves encrypting data about transactions and sequentially sharing—as in creating a chain—the record across a network or block of computers.[14] The records are decentralized, duplicated, and distributed across computers on the internet. Each keeps a tally of the chain as proof of the original while also recording changes. How all this really works is so difficult to explain that every blockchain developer prefers to call it a "digital ledger." You would be hard pressed to find any narrative about blockchains that doesn't resort to this analogy. Such analogical leaps are so common in computing that it is easy to lose focus of what is truly occurring within the technology.

On top of the obvious complexity, many computing processes are fuzzy—meaning they cannot be understood even by the individuals who write the program. This makes computing processes appear even more mysterious. Take the case of artificial intelligence, by which I mean the different manifestations of machine learning algorithms that are applied to analyze data about user behavior collected by search engines, social media, and apps. Machine learning algorithms churn through vast datasets, looking for patterns. The algorithms learn what to look for, which is heavily influenced by the data available, what the analysis is searching for, and how the data was combined. There is no theory guiding this process, so while the final output may be interpretable, the analysis isn't. It is inexplicable, even to AI designers. This fuzziness adds to the bias that computing processes are inherently sophisticated.

The third reason the sophistication bias is common in cybersecurity is because of something we already discussed in chapter 3—that the field is dominated by engineers who love to focus on computing technology, its hardware, and its software. Engineers like to look for solutions where they are comfortable looking for them—not necessarily where the problems are. It's the streetlight effect.

The fourth reason is that no one wants to admit they were easily hacked, especially not security and technology corporations, which are expected to know better and be better than the average Main Street tech shop. Many security firms are worth hundreds of millions of dollars, and their market capitalization is at stake.

Most are also run by men, who dominate the field. In fact, 8 of 10 cybersecurity professionals are males.[15] Many don't like to be embarrassed into admitting when they are tricked by a teen using a simple attack vector. Saying it is a sophisticated attack, and one by a nation, removes the burden of admitting that their practices were flawed. It makes it easier to pass the blame, making the call for more-sophisticated countermeasures that much easier.

This leads to the last reason the sophistication bias is endemic in cybersecurity—that it helps sell more computing and security products. Having a virtual arms race is worth billions more than simple fixes such as conducting a CRS-based risk assessment or a cyber hygiene assessment using the CHI. According to estimates by Gartner Inc., by 2026, AI augmentation is expected to create $2.9 trillion in business value, and blockchain technology is expected to grow to over $360 billion.[16] No one wants to miss out on this growth—not technology companies, not investors, and surely not the media, which benefits from promoting such news. They get to report on new product launches and present reviews of the next highly sophisticated "magic bullet" that will solve all our security woes.

The sophistication bias in cybersecurity is a self-serving loop, and it doesn't just impact our understanding of the causes of the event. It extends to the solutions we think about as well. Because of the perceived sophistication of the problem, we expect the solution to be commensurately complex, but do complex problems always require sophisticated solutions? Paradoxically, they don't.

Take COVID-19, the pandemic that began in December 2019. It was caused by an updated version of a virus that had been identified in 2003 during another outbreak, severe acute respiratory syndrome (SARS), which had infected 8,500 people worldwide and killed about 495.[17] The 2019 variant, named SARS-COV-2, was an even bigger killer.[18] By the end of 2021, it had infected 285 million people and killed about 5.5 million all over the world.[19] It had decimated lives, families, economies, and nations.

The reason for this devastation is the sophistication of coronaviruses. They are among the few RNA viruses with a genomic proofreading mechanism. This protects coronaviruses against mutations that could weaken them. It also helps them resist antiviral drugs. Adding to this, SARS-COV-2 had special receptors that helped it latch onto the lower respiratory system. One virologist estimated it had a one hundred to one thousand times greater chance than the SARS virus of getting deep into our lung.[20] Given its sophistication, it required an mRNA vaccine that could teach the body's cells to look for the same protein found in COVID-19 and destroy it.[21] Creating this was a highly complex endeavor. The pharmaceutical industry had to deploy enormous resources, involving thousands of human hours, multiple clinical trials, support from governments, and cutting-edge technology. The vaccine needed special storage at very low temperatures and had to be administered in multiple doses. Its production, transportation, handling, and distribution had to be perfectly choreographed. The scale of the operation was unparalleled in human history and it cost billions of dollars.

But even as the world held its collective breath and waited for a vaccine, there was a simple solution. It involved wearing a mask, washing hands, and avoiding public gatherings. It was a low-technology solution that worked to stall the disease, even kill its progress. The same "technology" had been successful a century earlier, during another pandemic. Beginning around 1918, a highly contagious influenza pandemic, which came to be called the Spanish Flu pandemic, spread around the world. It killed an estimated 20 million to 50 million people, roughly 3 percent of the world's population. There was no vaccine for it and no cure. The pandemic's toll was choked using the same low-tech solutions we resorted to with COVID-19. People were asked to wear masks, avoid public places, avoid shaking hands, and to stay indoors.[22]

Our response to COVID-19 and the Spanish Flu shows how complex problems can be effectively stalled and even solved with simple solutions. The type of solution that works in any given situation is not connected to the complexity of the problem. Often there is no parallel between them. Yet, in cybersecurity, after every major breach, we focus on its sophistication and look for increasingly sophisticated solutions. The sophistication bias is the reason why five years after the Sony hack and after spending billions of dollars worldwide on various solutions, attacks like SolarWinds still occur. As Albert Einstein famously said, we cannot solve our problems with the same thinking we used to create them. We need to think differently.

This is what the Cyber Risk Survey process and the Cyber Hygiene Inventory offer (figure 10.1). Rather than solve cybersecurity problems using the technology that created them, these tools focus on the targets

**Figure 10.1**
The advantages of the Cyber Risk Survey and Cyber Hygiene Inventory.

of hackers—users. They help diagnose the user and understand who is at risk, how much risk, and why users are at risk. By using an evidence-based approach that is agnostic to operating systems or software, they help build solutions around the problem.

In a vendor-driven IT solution space full of jargon, the CRS provides simplicity. In place of obscure, proprietary approaches and sophisticated technical fixes, it provides a transparent, actionable measure; in place of top-down solutions that prejudge the problem, it provides a process to derive an evidence-based diagnosis; and in place of pass-fail metrics that lack validity, it returns a simple individual risk assessment. It helps create pen tests that are valid and computes an organization-wide net resilience score (NRS). The CRS achieves all this without needing to license new software, bring in more technology, or impose any restrictions on users. The CHI supplements these efforts by providing an understanding of what users know, understand, do, and can do to achieve cyber resilience.

As the case studies in chapter 9 demonstrate, the evidence-based process the CRS and CHI create can help organizational IT not just understand why users are at risk from social engineering but also discover, implement, and track the viability of solutions that may not require more technology. Best of all, unlike the unproven promises of tomorrow's technologies, the approaches presented in this book are available to you today.

You and every person working in IT to solve the problem of social engineering can implement them right away. With these tools, organizations can tilt the information advantage, the knowledge environment, the information environment, and the motivational advantages away from hackers and back in their favor.

## REVERSING THE HACKER'S INFORMATION ADVANTAGE

The social engineer's information advantage comes from having access to privileged user information from prior attacks. Pretexting and spear phishing are both the sources and conduits for procuring and weaponizing such information.

The CRS approach addresses this at three levels. First, it allows IT departments to identify who is likely to fall victim to attacks. This helps create information-based solutions and early warning systems to protect the high-risk or likely-to-be-targeted users. For instance, alerts can be placed informing IT staff whenever information about such users is available on the Dark Web. The network access provided to such individuals can also be curtailed and monitored more closely. This way, IT managers know about the threat before these users are even targeted, so extra protections can be put in place.

Second, once the highest-risk users are identified, IT departments can build multiple layers of technological defenses around them. This forces hackers, who reuse attacks and apply the same attack across multiple organizations, to craft attacks that are specific to each organization and its users. This not only renders any broad attack less likely to succeed but also increases the effort cost of the attack. It renders the organization a less attractive target.

Third, the CRS allows IT staff to understand not just the likely targets of an attack but also the likely attacks that users might need to be prepared against. Here the CRS benchmarking process helps. Using a handful of representative users, it provides a consensus score of the potency of any attack. Using the V-Triad and the SCAM framework, the score and the reasons for it can be understood, so if a new form of attack is in the news, the potency of the attack on the users in the organization can be determined in advance. The approach can be expanded to score Wi-Fi spoofing, USB drop-offs, and other forms of social engineering attacks. Using this, IT departments can determine whether an attack requires an urgent email announcement or any other countermeasures. The countermeasures can be technological or, again using the CRS risk data, focused solely at the weakest links. In this way, organizations can prepare for current, ongoing, and future attacks. The information advantage of social engineers can be checked by implementing the CRS.

### Reversing the Hacker's Knowledge Advantage

The social engineer's knowledge advantage comes from hackers' access to better technical resources. The CRS allows three levels of knowledge that reduce the advantages of the social engineer: meta, micro, and macro. Layered on this, the CHI provides data on what security behaviors users are aware of doing, can do, know to do, and actually do. Together, they provide deep insights into users.

First, the CRS provides metaknowledge about why users fall victim to social engineering. This is unlike any other existing training approaches, which presuppose knowledge gaps in users. The data derived from the CRS can be arrayed against the CHI score of each user to pinpoint the specific security thought or behavior that needs to be addressed. By knowing the why and the specific what, IT departments can address the issues in advance of being exploited.

Second, the CRS categorizes users based on how much risk they pose to the enterprise. This stratification provides micro-level knowledge of different groups and their associated risks. It allows IT departments to deal with individuals and risky groups by using different approaches. They can adjust their policies based on need and evidence and rely less on blanket approaches that are applied across the organization. The specific policy and best-practice suggestions can be defined by the relative CHI scores, so users from various at-risk groups who need different hygiene interventions can be provided with them.

Third, the NRS provides an aggregated risk score that can be cumulated to provide macro data. The CHI provides aggregated hygiene scores that provide a snapshot of the level of resilience of a group or a division within the organization and where there is a lack. They can be tracked over time to see improvements and assess the relative values of different interventions. For organizations such as insurers and for policy makers interested in understanding user risk at higher-order levels, the data can be pooled to provide sector-level insights.

The three levels of knowledge obtained through the CRS and CHI cover fundamental user-level processes such as cognition and behavior. They are technology, device, software application, and operating system agnostic. This makes the knowledge applicable across organizations, technologies, attack types, and even time. Thus, on multiple levels, the knowledge obtained using the CRS along with the CHI defeats the knowledge advantages of the social engineer.

### Changing the Information Environment and Incentive Framework

Today's information environment fosters secrecy rather than sharing. This includes information about breaches, pen tests, their failure rates, and about interventions and policies that organizations have instituted. Because of this, most organizations continually reinvent the proverbial security wheel.

One reason for this is that at the user level the pen-test data is all these organizations really have, so sharing it would critically expose them. In place of this, the CRS and CHI provide a range of metrics, some of which can be shared without jeopardizing the entire organization. For instance, organizations can share the NRS or even their ongoing pen-test failure data without revealing the meta- or microknowledge gleaned from the CRS.

They could release CHI scores for awareness and knowledge, without providing information about capacity. Being able to share information more widely leads not just to greater transparency but also to greater accountability. This way, organizations that suffer a breach are held liable for just what they knowingly ignored, not for what was beyond their capacity to control.

It is also currently impossible to audit organizations to understand their true level of preparedness. All that a federal regulator or insurance provider currently has is the pen-test data provided by organizations, which, as we have already discussed, utilizes pen tests of varying and unknown quality. It would be foolhardy to define policies and insurance premiums based on these estimates.

Using the CRS benchmarking process, a regulatory agency can develop a series of valid tests and, using representative users in different organizations within a sector, derive a baseline score for the attack. IT staff in the organizations can then be tasked with deploying these tests using the CRS. The results would then be comparable across the organizations. The same process can locate solutions organically, in the open-ended CRS responses submitted by users. Regulators can reduce their reliance on cybersecurity consultancies by using such metrics and solutions. They can move from being just an auditor to becoming a solution provider for the organizations under their purview.

Today, organizational IT is focused on repeated testing to increase awareness, assess risk, and improve readiness. The lack of a unifying framework guiding any of this has prompted the implementation of even stricter security restrictions and, invariably, even more training. This has led to training fatigue, users learning from the test, users finding workarounds, and managers informing users of tests—all to reduce phishing failure rates. This reduces accountability both from the users subjected to them and the organizations using them.

The CRS approach doesn't require repeated testing, because it uses a valid pen-test development and measurement framework, providing reliable data using far fewer tests. The same is true for the CHI. It can be used just once or twice a year to assess and track users' cybersecurity readiness.

Finally, and perhaps most importantly, the goal of CRS isn't to drive down a single statistic but rather to understand users. Designed as a diagnostic tool, it is an evidence-based process for finding actionable solutions to address users' security vulnerabilities. The motivation for implementing it is not to trick users into falling for a phishing pen test but rather to understand who might and why. Along with the CHI, it is built to explain what in attacks and users causes vulnerability and how this can be remediated. This creates accountability—a critical requirement for fostering change—and ensures that those who are vulnerable are not just notified of their weakness but also taught how to fix it. In using these tools, the IT department's motivation shifts away from looking for the keys where the light shines to shining a light where the keys can be found.

## THE CHANGING ROLE OF IT MANAGEMENT

Throughout this book, we've discussed the changing role of IT management in today's corporations. Gone are the days when IT departments could control what devices users brought into organizations and what they could do on them. Despite this, the responsibility for breaches continues to be placed on the shoulders of IT managers.

The pressure is real, made all the more so more by varying guidelines, poorly devised policies, and an engineering-oriented, top-down approach to users. The CRS and CHI convert the role of IT managers into agents of change. They are no longer just engineers implementing a solution or police officers disciplining users. They are instead problem solvers who apply cognitive and behavioral science to solve the people problem of social engineering.

We began in chapter 1 with the story of the Thuggee cult, among the earliest social engineers, and the story of Henry Sleeman, who, through the application of innovative scientific techniques of the time, collected profiling data on the cult's members and their likely victims. He worked to understand the users and the likely targets of the cult. He had policemen disguise themselves as travelers and join caravans to protect those most likely to be targeted. This helped him apprehend cult members, but he had to do more to destroy them. He had to influence how people—travelers, villagers, townsfolk, chieftains, and local heads of state—thought of the cult. For this, he had to become a change agent.

We have lost to history much of how Sleeman accomplished this, but thanks to the many books he authored, we can piece together his steps. We know he began by developing a rapport with the people. He learned local languages, traveled extensively, and studied the history, zoology, and demography of the people. (His writings about wild children raised by wolves in jungles inspired Rudyard Kipling to write *The Jungle Book*.) He worked to *build a relationship of trust* so he could learn about the people and understand their support of the cult. This was necessary to get to the next pivotal step: *diagnosing the problem.*

Sleeman then figured out the core problem, that Thuggees had many benefactors, solidified by marriage and alliances. To reverse this tradition, he began identifying the local groups that were most likely to support the cult and those that didn't. From the latter he had to *find champions* who helped foster an *intent for change* in the community. He convinced some to infiltrate the cult, others to join the police force, and still others to report on suspicious activities they observed. Sleeman rewarded them with salaries, titles, commendations, and social recognition. This convinced many others and helped convert *intent into action*. Sleeman also *identified the weak links* needing protection and built defenses around them. Finally, he developed policies and processes to *stabilize adoption* of his novel approaches, to ensure that progress continued and that no other group arose to fill the vacuum. For this, he passed new laws outlawing thuggery, established the Thuggee and Dacoity Department, and through them institutionalized his practices. This helped foster continued participation from the population and choked the centuries-old scourge from ever gaining a foothold again.

The steps Sleeman took are exactly what business school textbooks today teach executives interested in creating change in organizations.[23] What worked then works just as well today.

While the Thuggees of history are long dead, their analogues in the cyber world thrive. They appear as hacktivists, organized gangs, and hackers. Many attack in unison, coordinating their efforts, using pseudonyms and secret codes, and sharing their stolen wares in dark marketplaces. Their primary targets are the users, particularly the weakest ones, through whom they enter organizations and pillage.

Standing in the modern social engineer's path is you—the IT manager, the CISO, the consultant, the security specialist, the business owner. You need to do what Sleeman did: establish trust with the users, diagnose the problem, identify the weak links, create an intent for change, convert intent into action, and stabilize adoption. You need to apply the science of measurement, data gathering, and analysis to profile users, find the weakest links, and protect them. And now you can.

# Notes

**INTRODUCTION**

1. Peter Elkind, "Sony Pictures: Inside the Hack of the Century," *Fortune*, June 25, 2015, https://fortune.com/longform/sony-hack-part-1/.

2. Arjun Kharpal, "North Korea Accidentally Lets World Access Its Internet and It Only Has 28 Websites," *CNBC*, September 21, 2016, https://www.cnbc.com/2016/09/21/north-korea-accidentally-lets-world-access-its-internet-and-it-only-has-28-websites.html.

3. Central Intelligence Agency, *The World Factbook*, accessed December 27, 2020, https://www.cia.gov/library/publications/resources/the-world-factbook/geos/kn.html.

4. Brenden I. Koerner, "Inside the Cyberattack That Shocked the US Government," *Wired*, October 23, 2016, https://www.wired.com/2016/10/inside-cyberattack-shocked-us-government/.

5. Tom Lamont, "Life after the Ashley Madison Affair," *The Guardian*, February 27, 2016, https://www.theguardian.com/technology/2016/feb/28/what-happened-after-ashley-madison-was-hacked.

6. Verizon, *2017 Verizon Data Breach Investigations Report*, 2017, https://www.knowbe4.com/hubfs/rp_DBIR_2017_Report_execsummary_en_xg.pdf.

7. FireEye, *MTrends Report*, 2018, https://www.fireeye.com/current-threats/annual-threat-report/mtrends.html.

8. Frank R. Konkel, "Pentagon Thwarts 36 Million Email Breach Attempts Daily," *NextGov*, January 11, 2018, https://www.netgov.com/cybersecurity/2018/01/pentagon-thwarts-36-million-email-breach-attempts-daily/145149/.

9. Deann D. Caputo, Shari Lawrence Pfleeger, Jesse D. Freeman, and M. Eric Johnson, "Going Spear Phishing: Exploring Embedded Training and Awareness," *IEEE Security & Privacy* 12, no. 1 (2013): 28–38; Aaron J. Ferguson, "Fostering E-Mail Security Awareness: The West Point Carronade," *Educause Quarterly* 28, no. 1 (2005): 54–57.

10. Dennis K. Berman, "Adm. Michael Rogers on the Prospect of a Digital Pearl Harbor," *Wall Street Journal*, October 26, 2015, http://www.wsj.com/article_email/adm-michael-rogers-on-the-prospect-of-a-digital-pearl-harbor-1445911336-lMyQjAxMTA1NzIxNzcyNzcwWj.

11. Arun Vishwanath, "Why Most Cyber Security Training Fails and What We Can Do about It," Black Hat, November 29, 2017, https://www.youtube.com/watch?v=3L3IrAN30a4 &feature=emb_title.

12. Michelle P. Steves, K. Kristen Greene, and Mary F. Theofanos, "Categorizing Human Phishing Detection Difficulty: A Phish Scale, *Journal of Cybersecurity* 6, no. 1 (2020); doi: 10.1093/ cybsec/tyaa009.

## CHAPTER 1

1. Joseph M. Hatfield, "Social Engineering in Cybersecurity: The Evolution of a Concept," *Computers & Security* 73 (2018): 102–113.

2. John Gray, *An Efficient Remedy for the Distress of Nations* (Edinburgh: A. and C. Black, 1842).

3. Ian D. Wyatt and Daniel E. Hecker, "Occupational Changes during the 20th Century," *Monthly Labor Review*, March 2006, 35–57, https://www.bls.gov/mlr/2006/03/art3full.pdf.

4. Chris Baraniuk, "Whatever Happened to the Phone Phreakers?," *The Atlantic*, February 20, 2013, https://www.theatlantic.com/technology/archive/2013/02/whatever-happened -to-the-phone-phreaks/273332/.

5. Phil Lapsley, *Exploding the Phone: The Untold Story of the Teenagers and Outlaws Who Hacked Ma Bell* (New York: Grove Press, 2013).

6. Steven V. Brull, "At 7 1/2 Cents a Minute, Who Cares If You Can't Hear a Pin Drop?," *Business Week*, December 29, 1997, https://www.bloomberg.com/news/articles/1997-12-28/at -7-1-2-cents-a-minute-who-cares-if-you-cant-hear-a-pin-drop.

7. Lapsley, *Exploding the Phone*.

8. Mark Thomson, "How Disbanding the Iraqi Army Fueled ISIS," *Time*, May 28, 2015, https://time.com/3900753/isis-iraq-syria-army-united-states-military/.

9. Robert M. Pirsig, *Zen and the Art of Motorcycle Maintenance: An Inquiry into Values* (New York: William Morrow, 1999).

10. Erin Fuchs, "Here's the 1898 Version of Those Nigerian Email Scams," *Business Insider*, July 14, 2014, https://www.businessinsider.com/charles-seife-writes-about-the-origin-of -the-spanish-prisoner-scam-2014-7.

11. Finn Brunton, *Spam: A Shadow History of the Internet* (Cambridge, MA: MIT Press, 2013).

12. Finn Brunton, "The Long, Weird History of the Nigerian Email Scam," *Boston Globe*, May 19, 2013, https://www.bostonglobe.com/ideas/2013/05/18/the-long-weird-history-nigerian-mail -scam/C8bIhwQSVoygYtrlxsJTlJ/story.html.

13. Brunton, "History of the Nigerian Email Scam."

14. Office of the Inspector General, United States Postal Service, *Semiannual Report on the Audit, Investigative, and Security Activities of the United States Postal Service, April 1–September 30, 2009*, accessed January 4, 2021, https://www.uspsoig.gov/sites/default/files/document -library-files/2015/fall09.pdf.

15. "The Ethics of Scambaiting," 419eater, accessed January 4, 2021, https://www.419eater .com/html/ethics.htm.

16. Christine Hauser, "U.S. Breaks Up Vast I.R.S. Phone Scam," *New York Times*, July 23, 2018, https://www.nytimes.com/2018/07/23/business/irs-phone-scams-jeff-sessions.html.

17. Federal Bureau of Investigation, Internet Crime Complaint Center, *2017 Internet Crime Report*, 2017, https://pdf.ic3.gov/2017_IC3Report.pdf.

18. Brunton, "History of the Nigerian Email Scam."

19. "History of Phishing," Phishing.org, 2020, http://www.phishing.org/history-of-phishing.

20. Lily Hay Newman, "Nigerian Email Scammers Are More Effective Than Ever," *Wired*, May 3, 2018, https://www.wired.com/story/nigerian-email-scammers-more-effective-than-ever/.

21. Jack Stubbs, Raphael Satter, and Christopher Bing, "Exclusive: Obscure Indian Cyber Firm Spied on Politicians, Investors Worldwide," Reuters, June 9, 2020, https://www.reuters .com/article/us-india-cyber-mercenaries-exclusive-idUSKBN23G1GQ.

22. Jordan Valinsky, "Shark Tank Host Loses $400,000 in a Scam," *CNN*, February 27, 2020, https://www.cnn.com/2020/02/27/business/barbara-corcoran-email-hack-trnd/index .html.

23. Thomas Brewster, "Hacker Pretends to Be Evan Spiegel to Steal Snapchat Employee Data," *Forbes*, February 29, 2016, https://www.forbes.com/sites/thomasbrewster/2016/02/29 /snapchat-data-leak/?sh=2e5bb6b33176.

24. Tom Huddleston Jr., "How This Scammer Used Phishing Emails to Steal over $100 Million from Google and Facebook," *CNBC*, March 27, 2019, https://www.cnbc.com/2019 /03/27/phishing-email-scam-stole-100-million-from-facebook-and-google.html.

25. "North Korea Stole $2bn for Weapons via Cyber-attacks," *BBC*, August 7, 2019, https:// www.bbc.com/news/world-asia-49259302.

26. "APT38: Unusual Suspects," FireEye, 2018, https://vision.fireeye.com/editions/01/north -koreas-unusual-suspects.html.

27. Verizon, *2018 Verizon Data Breach Investigations Report*, 2018, https://enterprise.verizon .com/resources/reports/DBIR_2018_Report.pdf.

28. FireEye-Mandiant, *MTrends 2018*, 2018, https://www.fireeye.com/content/dam/collateral /en/mtrends-2018.pdf.

**CHAPTER 2**

1. Raphael Satter, Jeff Donn, and Chad Day, "Inside Story: How Russians Hacked the Democrats' Emails," Associated Press, November 4, 2017, https://apnews.com/article/dea73efc0 1594839957c3c9a6c962b8a.

2. Rick McCormick, "Hack Leaks Hundreds of Nude Celebrity Photos," *The Verge*, September 1, 2014, https://www.theverge.com/2014/9/1/6092089/nude-celebrity-hack.

3. Matt MacInnis, "How I Survived and Thrived in Apple's Legendary Environment of Super-Secrecy," *Vox*, September 11, 2017, https://www.vox.com/2017/9/11/16288896/apple-secrecy-inkling-culture-leadership-transparency-values.

4. Evan Perez, "Sources: US Officials Warned DNC of Hack Months before the Party Acted," *CNN*, July 26, 2016, https://www.cnn.com/2016/07/25/politics/democratic-convention-dnc-emails-russia/index.html.

5. "The Phishing Email That Hacked the Account of John Podesta," *CBS News*, October 28, 2016, https://www.cbsnews.com/news/the-phishing-email-that-hacked-the-account-of-john-podesta/.

6. Taylor Covington, "Burglary Statistics, Research, and Facts," The Zebra, October 2, 2020, https://www.thezebra.com/resources/research/burglary-statistics/.

7. Justin Ling, Rachel Browne, and Tamara Khandaker, "Meet the 22-Year-Old Canadian Luxury Car Aficionado the FBI Says Helped the Russian Yahoo Hackers," *Vice News*, March 15, 2017, https://news.vice.com/en_us/article/9kd7ea/meet-the-22-year-old-canadian-luxury-car-aficionado-the-fbi-says-helped-the-russian-yahoo-hackers.

8. Erin Pearson, "No Jail for Teen Who Hacked His Way into Apple's Secure Systems," *The Age*, September 27, 2018, https://www.theage.com.au/national/victoria/no-jail-for-teen-who-hacked-his-way-into-apple-s-secure-systems-20180927-p506ca.html.

9. "Read Mueller Probe Indictment of 12 Russians for Hacking Democrats," *Washington Post*, March 22, 2019, https://www.washingtonpost.com/context/read-mueller-probe-indictment-of-12-russians-for-hacking-democrats/d72bca11-93cf-40e2-87ec-ea6ab02f1dcf/.

10. Russ Read, "How a Handful of Russian Techies Pulled One of the Biggest Online Heists in History," *Daily Caller News Foundation*, March 15, 2017, https://dailycaller.com/2017/03/15/how-4-russian-cyber-spies-pulled-off-one-of-the-biggest-cyber-heists-in-history/.

11. Sheera Frenkel, Nathaniel Popper, Kate Conger, and David E. Sanger, "A Brazen Online Attack Targets V.I.P. Twitter Users in a Bitcoin Scam," *New York Times*, July 15, 2020, https://www.nytimes.com/2020/07/15/technology/twitter-hack-bill-gates-elon-musk.html.

12. Marcus Gilmer, "Teen Who Hacked Apple Told to Use 'Gifts for Good Rather Than Evil,'" *Mashable*, May 27, 2019, https://mashable.com/article/australia-teen-apple-hack/.

13. "The Morris Worm: 30 Years since First Major Attack on Internet," FBI, November 2, 2018, https://www.fbi.gov/news/stories/morris-worm-30-years-since-first-major-attack-on-internet-110218.

14. Andrada Fiscutean, "Hunting Vintage MS-DOS Viruses from Cuba to Pakistan," CSO Online, July 24, 2019, https://www.csoonline.com/article/3409790/hunting-vintage-ms-dos-viruses-from-cuba-to-pakistan.html.

15. A. Jayanthi, "First Known Ransomware Attack in 1989 Also Targeted Healthcare," *Beckers Hospital Review*, May 11, 2016, https://www.beckershospitalreview.com/healthcare-information-technology/first-known-ransomware-attack-in-1989-also-targeted-healthcare.html.

16. "Microsoft Acquires Hotmail," Microsoft, December 31, 1997, https://news.microsoft.com/1997/12/31/microsoft-acquires-hotmail/.

17. "Press Release: Google to Acquire DoubleClick," *Wall Street Journal*, April 14, 2007, https://www.wsj.com/articles/SB117649971255169544.

18. Jesse Damiani, "Your Social Security Number Costs $4 on the Dark Web, New Report Finds," *Forbes*, March 25, 2020, https://www.forbes.com/sites/jessedamiani/2020/03/25/your-social-security-number-costs-4-on-the-dark-web-new-report-finds/?sh=7e0bbc1b13f1.

19. Peter Goffin, "Ex-classmates of Karim Baratov Recall Wealthy Introvert with Flashy Cars, Expensive Clothes," *Hamilton Spectator*, March 16, 2017, https://www.thespec.com/news-story/7191984-ex-classmates-of-karim-baratov-recall-wealthy-introvert-with-flashy-cars-expensive-clothes/.

20. Robert McMillan, "The World's First Computer Password? It Was Useless Too," *Wired*, January 27, 2012, https://www.wired.com/2012/01/computer-password/.

21. Matt Weinberger, "The Rise of Bill Gates, from Harvard Dropout to Richest Man in the World," *Business Insider*, December 26, 2017, https://www.businessinsider.com/the-rise-of-bill-gates-2016-3.

22. Dancho Danchev, "Malware Detected at the International Space Station," ZDNET, August 26, 2008, https://www.zdnet.com/article/malware-detected-at-the-international-space-station/.

23. FireEye-Mandiant, *M Trends 2018*, 2018, https://www.fireeye.com/content/dam/collateral/en/mtrends-2018.pdf.

24. Mike Felch, "Stealing 2FA Tokens on Red Teams with CredSniper," Black Hills Information Security, August 20, 2018, https://www.blackhillsinfosec.com/stealing-2fa-tokens-on-red-teams-with-credsniper/.

25. Kelly Sheridan, "Cracking 2FA: How It's Done and How to Stay Safe," Dark Reading, May 18, 2018, https://www.darkreading.com/endpoint/cracking-2fa-how-its-done-and-how-to-stay-safe/d/d-id/1331835.

26. Nick Hopkins, "Deloitte Hack Hit Server Containing Emails from across US Government," *The Guardian*, October 10, 2017, https://www.theguardian.com/business/2017/oct/10/deloitte-hack-hit-server-containing-emails-from-across-us-government.

27. Lily Hay Newman, "Microsoft Email Hack Shows the Lurking Danger of Customer Support," *Wired*, April 15, 2019, https://www.wired.com/story/microsoft-email-hack-outlook-hotmail-customer-support/.

28. Heather Murphy, "When a DNA Test Says You're a Younger Man, Who Lives 5,000 Miles Away," *New York Times*, December 7, 2019, https://www.nytimes.com/2019/12/07/us/dna-bone-marrow-transplant-crime-lab.html.

29. Stanley Milgram, "The Small World Problem," *Psychology Today* 2, no. 1 (1967): 60–67, http://snap.stanford.edu/class/cs224w-readings/milgram67smallworld.pdf.

30. Iain Thomson, "US Pentagon Scrambles after Strava Base Leaks. Here's a Summary of the New Rules: 'Secure That S***, Hudson!,'" *The Register*, January 29, 2018, https://www.theregister.co.uk/2018/01/29/us_pentagon_strava_tracking/.

31. Brian Krebs, "Sextortion Scam Uses Recipient's Hacked Passwords," Krebs on Security, July 12, 2018, https://krebsonsecurity.com/2018/07/sextortion-scam-uses-recipients-hacked-passwords/.

32. Anti-Phishing Working Group (APWG), *Phishing Activity Trends Report, 4th Quarter 2018*, 2019, https://docs.apwg.org/reports/apwg_trends_report_q4_2018.pdf; Marco Cova, Christopher Kruegel, and Giovanni Vigna, "There Is No Free Phish: An Analysis of "Free" and Live Phishing Kits," *WOOT* 8 (2008):1–8, https://www.usenix.org/legacy/event/woot08/tech/full_papers/cova/cova_html/.

33. Ellen Nakashima and Craig Timberg, "NSA Officials Worried about the Day Its Potent Hacking Tool Would Get Loose. Then It Did," *Washington Post*, May 16, 2017, https://www.washingtonpost.com/business/technology/nsa-officials-worried-about-the-day-its-potent-hacking-tool-would-get-loose-then-it-did/2017/05/16/50670b16-3978-11e7-a058-ddbb23c75d82_story.html.

34. Koen Van Impe, "How Can an ISAC Improve Cybersecurity and Resilience?," Security Intelligence, July 16, 2018, https://securityintelligence.com/how-can-an-isac-improve-cybersecurity-and-resilience/.

35. Georgina Torbet, "Baltimore Ransomware Attack Will Cost the City over $18 Million," Engadget, June 6, 2019, https://www.engadget.com/2019/06/06/baltimore-ransomware-18-million-damages/.

36. Ian Duncan and Christine Zhang, "Analysis of Ransomware Used in Baltimore Attack Indicates Hackers Needed 'Unfettered Access' to City Computers," *Baltimore Sun*, May 17, 2019, https://www.baltimoresun.com/politics/bs-md-ci-ransomware-attack-20190517-story.html.

37. William Burr, Donna Dodson, and W. Polk, *Electronic Authentication Guideline* (NIST Special Publication No. SP 800-63, Version 1.0 [withdrawn]) (Gaithersburg, MD: National Institute of Standards and Technology).

38. P. Grassi, Michael E. Garcia, and James L. Fenton, *Digital Identity Guidelines* (NIST Special Publication SP 800-63, Version 3) (Gaithersburg, MD: National Institute of Standards and Technology, 2017).

39. Isobel Koshiw, "How an International Hacker Network Turned Stolen Press Releases into $100 Million," *The Verge*, August 22, 2018, https://www.theverge.com/2018/8/22/17716622/sec-business-wire-hack-stolen-press-release-fraud-ukraine.

## CHAPTER 3

1. Scott Mayerowitz, "The Suitcase with Wheels Turns 40," *ABC News*, October 1, 2010, https://abcnews.go.com/Travel/suitcase-wheels-turns-40-radical-idea-now-travel/story?id=11779469.

2.  Thomas S. Kuhn, *The Structure of Scientific Revolutions* (Chicago: University of Chicago Press, 2012).

3.  Adrienne Mayor, *The First Fossil Hunters: Paleontology in Greek and Roman Times* (Princeton, NJ: Princeton University Press, 2001).

4.  John Noble Wilford, "Greek Myths: Not Necessarily Mythical," *New York Times*, July 4, 2000, https://www.nytimes.com/2000/07/04/science/greek-myths-not-necessarily-mythical.html.

5.  Leonard J. Shustek, "In His Own Words: Gary Kildall," Computer History Museum, August 2, 2016, https://www.computerhistory.org/atchm/in-his-own-words-gary-kildall/.

6.  Arun Vishwanath, "Spear Phishing Has Become Even More Dangerous," *CNN*, September 1, 2018, https://www.cnn.com/2018/09/01/opinions/spear-phishing-has-become-even-more-dangerous-opinion-vishwanath/index.html.

7.  Bernard Marr, "How Much Data Do We Create Every Day? The Mind Blowing Stats Everyone Should Read," *Forbes*, May 21, 2018, https://www.forbes.com/sites/bernardmarr/2018/05/21/how-much-data-do-we-create-every-day-the-mind-blowing-stats-everyone-should-read/#4020c24960ba.

8.  Nicolas Christin, Serge Egelman, Timothy Vidas, and Jens Grossklags, "It's All about the Benjamins: An Empirical Study on Incentivizing Users to Ignore Security Advice," in *Lecture Notes in Computer Science,* vol. 7035, *Financial Cryptography and Data Security,* ed. G. Danezis, 16–30 (Berlin: Springer, 2011).

9.  Christin et al., "It's All about the Benjamins."

10. Alessandro Oltramari, Diane S. Henshel, Mariana Cains, and Blaine Hoffman, "Towards a Human Factors Ontology for Cyber Security," in *Proceedings of the Tenth International Conference on Semantic Technology for Intelligence, Defense, and Security (STIDS)* 2015, 26–33.

11. Christopher D. Wickens, Sallie E. Gordon, Yili Liu, and J. Lee, *An Introduction to Human Factors Engineering*, vol. 2 (Upper Saddle River, NJ: Pearson Prentice Hall, 2004).

12. Mark Wilson, "FISMA and OPM Awareness and Training Requirements and Related NIST Guidelines" (PowerPoint presentation, March 12, 2007), https://csrc.nist.gov/CSRC/media/Presentations/FISMA-and-OPM-Awareness-and-Training-Requirements/images-media/fissea2007_fisma-omb-requirements-nist-guidelines_wilson.pdf.

13. "Uses and Benefits of the Framework," National Institute of Standards and Technology (NIST), accessed January 20, 2021, https://www.nist.gov/cyberframework/online-learning/uses-and-benefits-framework.

14. National Institute of Standards and Technology (NIST), *Framework for Improving Critical Infrastructure Cybersecurity*, version 1.1, 2018, https://nvlpubs.nist.gov/nistpubs/CSWP/NIST.CSWP.04162018.pdf.

15. IDG Communications, "Security Priorities and Investments Outlined in Latest IDG Research," IDG, July 31, 2019, https://www.globenewswire.com/news-release/2019/07/31/1894927/0/en/Security-Priorities-Investments-Outlined-in-Latest-IDG-Research.html.

16. Martin C. Libicki, Lillian Ablon, and Tim Webb, *The Defender's Dilemma: Charting a Course toward Cybersecurity* (Santa Monica, CA: RAND Corporation, 2015).

17. Taryn Oesch, "Why Funding for Cybersecurity Training Is Growing," Training Industry, July 9, 2019, https://trainingindustry.com/blog/compliance/why-funding-for-cybersecurity-training-is -growing/.

18. "The Cybersecurity 2020: This Is the Biggest Problem with Cybersecurity Research," *Washington Post*, April 18, 2019, https://www.washingtonpost.com/news/powerpost/paloma /the-cybersecurity-202/2019/04/18/the-cybersecurity-202-this-is-the-biggest-problem -with-cybersecurity-research/5cb7a231a7a0a46fd9222a47/.

19. Vindu Goel, "Facebook Tinkers with Users' Emotions in News Feed Experiments, Stirring Outcry," *New York Times*, June 29, 2014, https://www.nytimes.com/2014/06/30 /technology/facebook-tinkers-with-users-emotions-in-news-feed-experiment-stirring -outcry.html.

20. Aaron J. Ferguson, "Fostering E-mail Security Awareness: The West Point Carronade," *EDUCASE Quarterly* 28, no. 1 (2005): 54–57.

21. Deanna D. Caputo, Shari Lawrence Pfleeger, Jesse D. Freeman, and M. Eric Johnson, "Going Spear Phishing: Exploring Embedded Training and Awareness," *IEEE Security & Privacy* 12, no. 1 (2014): 28–38.

22. Caputo et al., "Going Spear Phishing."

23. Shahryar Baki and Rakesh Verma, "Sixteen Years of Phishing User Studies: What Have We Learned?," 2021, arXiv preprint arXiv:2109.04661.

## CHAPTER 4

1. Adrian Willings, "The Very Best Internet Optical Illusions Around: You Won't Believe Your Eyes," Pocket-lint, April 24, 2020, https://www.pocket-lint.com/apps/news/140473-best -internet-optical-illusions-you-won-t-believe-your-eyes.

2. Susana Martinez-Conde, Dave Conley, Hank Hine, Joan Kropf, Peter Tush, Andrea Ayala, and Stephen L. Macknik, "Marvels of Illusion: Illusion and Perception in the Art of Salvador Dali," *Frontiers in Human Neuroscience* 9 (2015): 496, https://www.ncbi.nlm.nih.gov /pmc/articles/PMC4586274/.

3. Trevor J. Cox, "Scraping Sounds and Disgusting Noises," *Applied Acoustics* 69, no. 12 (2008): 1195–1204; Oliver Grewe, Björn Katzur, Reinhard Kopiez, and Eckart Altenmüller, "Chills in Different Sensory Domains: Frisson Elicited by Acoustical, Visual, Tactile and Gustatory Stimuli," *Psychology of Music* 39, no. 2 (2011): 220–239.

4. Alexis C. Madrigal, "Things You Cannot Unsee (and What That Says about Your Brain), *The Atlantic*, May 5, 2014, https://www.theatlantic.com/technology/archive/2014/05/10 -things-you-cant-unsee-and-what-that-says-about-your-brain/361335/.

5. Nikhil Swaminathan, "Why Does the Brain Need So Much Power?," *Scientific American* 29, no. 4 (2008): 2998, https://www.scientificamerican.com/article/why-does-the-brain-need-s/.

6.   Shelly Chaiken, "The Heuristic Model of Persuasion," in *Social Influence: The Ontario Symposium*, vol. 5, ed. Mark P. Zanna, James M. Olson, and C. P. Herman, 3–39 (New York: Psychology Press, 1987).

7.   Serena Chen and Shelly Chaiken, "The Heuristic-Systematic Model in Its Broader Context," in *Dual-Process Theories in Social Psychology*, ed. S. Chaiken and Y. Trope, 73–96 (New York: Guilford Press, 1999).

8.   Rachna Dhamija, J. Doug Tygar, and Marti Hearst, "Why Phishing Works," in *Proceedings of the SIGCHI Conference on Human Factors in Computing Systems*, ed. Rebecca Grinter, Thoms Rodden, Paul Aoki, Ed Cutrell, Robin Jeffries, and Gary Olson, 581–590 (New York: Association for Computing Machinery, 2006).

9.   Robert L. Whitwell, A. David Milner, and Melvyn A. Goodale, "The Two Visual Systems Hypothesis: New Challenges and Insights from Visual Form Agnosic Patient DF," *Frontiers in Neurology* 5 (2014): 255, https://www.frontiersin.org/articles/10.3389/fneur.2014.00255/full.

10.   Arun Vishwanath, "Examining the Distinct Antecedents of E-mail Habits and Its Influence on the Outcomes of a Phishing Attack," *Journal of Computer-Mediated Communication* 20, no. 5 (2015): 570–584.

11.   Hongxiu Li, Yong Liu, Xiaoyu Xu, Jukka Heikkilä, and Hans van der Heijden, "Modeling Hedonic IS Continuance through the Uses and Gratifications Theory: An Empirical Study in Online Games," *Computers in Human Behavior* 48 (2015): 261–272.

12.   "FBI Dir. Robert Mueller Talks Cybersecurity," CNET, October 7, 2009, https://www.youtube.com/watch?v=M1PzM51JF5s&feature=youtu.be.

13.   Robert LaRose, Carolyn A. Lin, and Matthew S. Eastin, "Unregulated Internet Usage: Addiction, Habit, or Deficient Self-Regulation?," *Media Psychology* 5, no. 3 (2003): 225–253.

14.   Deanna D. Caputo, Shari Lawrence Pfleeger, Jesse D. Freeman, and M. Eric Johnson, "Going Spear Phishing: Exploring Embedded Training and Awareness," *IEEE Security & Privacy* 12, no. 1 (2014): 28–38.

**CHAPTER 5**

1.   National Highway Traffic Safety Administration, *Crash Factors in Intersection-Related Crashes: An On-Scene Perspective* (US Department of Transportation Publication No. HS 811-366) (Washington, DC: US Department of Transportation, 2010), https://crashstats.nhtsa.dot.gov/Api/Public/ViewPublication/811366.

2.   Daniel Cadzow and Justin Booth, *Redesigning the Scajaquada Expressway*, Partnership for the Public Good, February 8, 2016. https://ecommons.cornell.edu/bitstream/handle/1813/73298/Environment__Scajaquada_Expressway.pdf?sequence=1. See also Sarah Goodyear, "A Deadly Expressway in the Middle of an Olmsted Park," *Bloomberg*, June 25, 2015, https://www.bloomberg.com/news/articles/2015-06-25/the-deadly-scajaquada-expressway-slices-through-an-olmsted-park-in-buffalo.

3. "About SANS Institute," SANS, 2021, https://www.sans.org/about/.

4. Donna Jalbert Patalano, "Police Power and the Public Trust: Prescriptive Zoning through the Conflation of Two Ancient Doctrines," *Boston College Environmental Affairs Law Review* 28 (2000): 683–718.

5. Brynne Harrison, Elena Svetieva, and Arun Vishwanath, "Individual Processing of Phishing Emails: How Attention and Elaboration Protect against Phishing," *Online Information Review* 40, no. 2 (2016): 265–281.

6. Arun Vishwanath, Brynne Harrison, and Yu Jie Ng, "Suspicion, Cognition, and Automaticity Model of Phishing Susceptibility," *Communication Research* 45, no. 8 (2016): 1146–1166.

## CHAPTER 6

1. Grzegorz Bilo, Oscar Sala, Carlotta Perego, Andrea Faini, Lan Gao, Anna Głuszewska, Juan Eugenio Ochoa, Dario Pellegrini, Laura Maria Lonati, and Gianfranco Parati, "Impact of Cuff Positioning on Blood Pressure Measurement Accuracy: May a Specially Designed Cuff Make a Difference?," *Hypertension Research* 40, no. 6 (2017): 573–580, https://www.ncbi .nlm.nih.gov/pmc/articles/PMC5506235/. See also Ernest Privšek, Margareta Hellgren, Lennart Råstam, Ulf Lindblad, and Bledar Daka, "Epidemiological and Clinical Implications of Blood Pressure Measured in Seated versus Supine Position," *Medicine* 97, no. 31 (2018): e11603, https://www.ncbi.nlm.nih.gov/pmc/articles/PMC6081069/.

2. Arun Vishwanath, "Why Most Cyber Security Training Fails and What We Can Do about It," Black Hat, November 29, 2017, https://www.youtube.com/watch?v=3L3IrAN30a4 &feature=emb_title.

3. Shane Frederick, "Cognitive Reflection and Decision Making," *Journal of Economic Perspectives* 19, no. 4 (2005): 25–42.

4. Deanna D. Caputo, Shari Lawrence Pfleeger, Jesse D. Freeman, and M. Eric Johnson, "Going Spear Phishing: Exploring Embedded Training and Awareness," *IEEE Security & Privacy* 12, no. 1 (2014): 28–38.

5. Michelle P. Steves, K. Kristen Greene, and Mary F. Theofanos, "Categorizing Human Phishing Detection Difficulty: A Phish Scale," *Journal of Cybersecurity* 6, no. 1 (2020), doi: 10.1093/cybsec/tyaa009.

## CHAPTER 7

1. Frederick F. Reichheld, "The One Number You Need to Grow," *Harvard Business Review* 81, no. 12 (2003): 46–55, https://hbr.org/2003/12/the-one-number-you-need-to-grow.

## CHAPTER 8

1. European Union Agency for Network and Information Security (ENISA), *Review of Cyber Hygiene Practices*, December 2016, https://www.enisa.europa.eu/publications/cyber-hygiene /at_download/fullReport.

2. "Stages, Factors and Symptoms of Cyber Diseases," *Cybersecurity Is Cyber Health* (blog), H-X Technologies, June 1, 2020, https://www.h-x.technology/blog/cybersecurity-cyber-health.

3. Arun Vishwanath, "Stop Telling People to Take Those Cyber Hygiene Multivitamins," in *Prepared for Evolving Threats: The Role of Behavioural Sciences in Law Enforcement and Public Safety*, ed. Majeed Khader, Gabriel Ong, and Carolyn Misir, 225–240 (Singapore: World Scientific, 2020), https://www.worldscientific.com/worldscibooks/10.1142/11812.

4. William Burr, Donna Dodson, and W. Polk, *Electronic Authentication Guideline* (NIST Special Publication No. SP 800-63, Version 1.0 [withdrawn]) (Gaithersburg, MD: National Institute of Standards and Technology, 2004).

5. C. Hassold, "Have We Conditioned Web Users to Be Phished?," PhishLabs, November 2, 2017, https://info.phishlabs.com/blog/have-we-conditioned-web-users-to-be-phished.

6. Anti-Phishing Working Group (APWG), *Phishing Activity Trends Report: 3rd Quarter 2019*, 2019, https://docs.apwg.org/reports/apwg_trends_report_q3_2019.pdf.

7. Mark D. Hill, Jon Masters, Parthasarathy Ranganathan, Paul Turner, and John L. Hennessy, "On the Spectre and Meltdown Processor Security Vulnerabilities," *IEEE Micro* 39, no. 2 (2019): 9–19.

8. Yoongu Kim, Ross Daly, Jeremie Kim, Chris Fallin, Ji Hye Lee, Donghyuk Lee, Chris Wilkerson, Konrad Lai, and Onur Mutlu, "Flipping Bits in Memory without Accessing Them: An Experimental Study of DRAM Disturbance Errors," *ACM SIGARCH Computer Architecture News* 42, no. 3 (2016): 361–372, https://users.ece.cmu.edu/~yoonguk/papers/kim-isca14.pdf.

9. FireEye-Mandiant, *MTrends 2018*, 2018, https://www.fireeye.com/content/dam/collateral/en/mtrends-2018.pdf.

10. Ferris Jabr, "How Does the Flu Actually Kill People?," *Scientific American*, December 18, 2017, https://www.scientificamerican.com/article/how-does-the-flu-actually-kill-people/.

11. Brett Molina, "Most People Don't Wash Their Hands Correctly, USDA Study Finds," *USA Today*, June 29, 2018, https://www.usatoday.com/story/news/nation-now/2018/06/29/usda-study-most-people-dont-wash-their-hands-correctly/745048002/.

12. Erik van Vulpen, "What Is Organizational Development? A Complete Guide," AIHR Digital, accessed February 11, 2021, https://www.digitalhrtech.com/organizational-development/.

13. Sidney G. Winter, "The Satisficing Principle in Capability Learning," *Strategic Management Journal* 21, no. 10–11 (2000): 981–996.

14. Arun Vishwanath, Loo Seng Neo, Pamela Goh, Seyoung Lee, Majeed Khader, Gabriel Ong, and Jeffery Chin, "Cyber Hygiene: The Concept, Its Measure, and Its Initial Tests," *Decision Support Systems* 128 (2020): 113160.

15. Vishwanath et al., "Cyber Hygiene."

## CHAPTER 10

1. David H. Freedman, "Why Scientific Studies Are So Often Wrong: The Streetlight Effect," *Discover Magazine*, December 9, 2010, https://www.discovermagazine.com/the-sciences/why-scientific-studies-are-so-often-wrong-the-streetlight-effect.

2. Arun Vishwanath, "Is 2016 the Year of Online Extortion?," *CNN*, March 25, 2016, https://www.cnn.com/2016/03/25/opinions/preventing-ransomware-attacks-vishwanath/index.html.

3. "The State of Ransomware in 2020," Blackfog, accessed February 18, 2021, https://www.blackfog.com/the-state-of-ransomware-in-2020/.

4. Daron Horwitz, "CYA Culture and the Importance of Admitting Mistakes," *Forbes*, February 28, 2016, https://www.forbes.com/sites/daronhorwitz/2016/02/28/cya-culture-and-the-importance-of-admitting-mistakes/?sh=697af75e14a6.

5. S. Kirchgaessner, "Jeff Bezos Hack: Amazon Boss's Phone 'Hacked' by Saudi Crown Prince," *The Guardian*, January 22, 2020, https://www.theguardian.com/technology/2020/jan/21/amazon-boss-jeff-bezoss-phone-hacked-by-saudi-crown-prince.

6. Colin Lecher and Russell Brandom, "Is Huawei a Security Threat? Seven Experts Weigh In," *The Verge*, March 17, 2019, https://www.theverge.com/2019/3/17/18264283/huawei-security-threat-experts-china-spying-5g.

7. Rajani Vaidyanathan, "Confessions of a Call-Centre Scammer," *BBC*, March 8, 2020, https://www.bbc.com/news/stories-51753362.

8. Vindu Goel and Suhasini Raj, "That Virus Alert on Your Computer? Scammers in India May Be behind It," *New York Times*, November 28, 2018, https://www.nytimes.com/2018/11/28/technology/scams-india-call-center-raids.html; Najmeh Miramirkhani, Oleksii Starov, and Nick Nikiforakis, "Dial One for Scam: A Large-Scale Analysis of Technical Support Scams," 2016, arXiv preprint arXiv:1607.06891.

9. Julien Legrand, "Artificial Intelligence as a Security Solution and Weaponizations by Hackers," *CISOMAG*, December 9, 2019, https://www.cisomag.com/hackers-using-ai/ .

10. S. Goldberg, "Isabella Stewart Gardner Museum Theft Audio Walk," Isabella Stewart Gardner Museum, 2020, https://www.gardnermuseum.org/sites/default/files/uploads/files/TheftAudioWalkTranscript_FINAL_20200301.pdf.

11. Jon Porter, "White House Now Says 100 Companies Hit by SolarWinds Hack, but More May Be Impacted," *The Verge*, February 18, 2021, https://www.theverge.com/2021/2/18/22288961/solarwinds-hack-100-companies-9-federal-agencies.

12. Bill Whitaker, "Solarwinds: How Russian Spies Hacked the Justice, State, Treasury, Energy and Commerce Departments," *60 Minutes*, CBS News, February 14, 2021, https://www.cbsnews.com/news/solarwinds-hack-russia-cyberattack-60-minutes-2021-02-14/.

13. Ryan Gallagher, "SolarWinds Adviser Warned of Lax Security Years before Hack," *Bloomberg*, December 21, 2020, https://www.bloomberg.com/news/articles/2020-12-21/solarwinds-adviser-warned-of-lax-security-years-before-hack?sref=zFmdEBXN.

14. "What Is Blockchain?," IBM, accessed February 20, 2021, https://www.ibm.com /blockchain/what-is-blockchain.

15. Steve Morgan, "Women Represent 20 Percent of the Global Cybersecurity Workforce in 2019," *Cybercrime Magazine*, March 28, 2019, https://cybersecurityventures.com/women -in-cybersecurity/.

16. "Gartner Says AI Augmentation Will Create $2.9 Trillion of Business Value in 2021," Gartner Inc., August 5, 2019, https://www.gartner.com/en/newsroom/press-releases/2019 -08-05-gartner-says-ai-augmentation-will-create-2point9-trillion-of-business-value-in -2021; "The Future of Blockchain," York Solutions, accessed February 20, 2021, https:// yorksolutions.net/the-future-of-blockchain-technology/.

17. Robert Ross, "Estimates of SARS Death Rates Revised Upward," Center for Infectious Disease Research and Policy, May 7, 2003, https://www.cidrap.umn.edu/news-perspective /2003/05/estimates-sars-death-rates-revised-upward.

18. "SARS Outbreak Contained Worldwide," World Health Organization, July 5, 2003, https://www.who.int/news/item/05-07-2003-sars-outbreak-contained-worldwide.

19. Hannah Ritchie, Edouard Mathieu, Lucas Rodés-Guirao, Cameron Appel, Charlie Giattino, Esteban Ortiz-Ospina, Joe Hasell, Bobbie Macdonald, Saloni Dattani, and Max Roser, "Coronavirus Pandemic (COVID-19)—The Data," Our World in Data, https:// ourworldindata.org/coronavirus.

20. David Cyranoski, "Profile of a Killer: The Complex Biology Powering the Coronavirus Pandemic," *Nature*, May 4, 2020, https://www.nature.com/articles/d41586-020-01315-7.

21. "Understanding mRNA Vaccines," Centers for Disease Control and Prevention, December 18, 2020, https://www.cdc.gov/coronavirus/2019-ncov/vaccines/different-vaccines/mrna .html.

22. Editors, "Spanish Flu," History, May 19, 2020, https://www.history.com/topics/world-war -i/1918-flu-pandemic.

23. Everett M. Rogers, *Diffusion of Innovations* (New York: Simon and Schuster, 2010).